EAST ASIAN HISTORICAL MONOGRAPHS

General Editor: WANG GUNGWU

Democracy Shelved:
Great Britain, China, and Attempts at
Constitutional Reform in Hong Kong,
1945–1952

EAST ASIAN HISTORICAL MONOGRAPHS

General Editor: WANG GUNGWU

The East Asian Historical Monographs series has, since its inception in the late 1960s, earned a reputation for the publication of works of innovative historical scholarship. It encouraged a generation of scholars of Asian history to go beyond Western activities in Asia seen from Western points of view. Their books included a wider range of Asian viewpoints and also reflected a stronger awareness of economic and socio-cultural factors in Asia which lay behind political events.

During its second decade the series has broadened to reflect the interest among historians in studying and reassessing Chinese history, and now includes important works on China and Hong Kong.

It is the hope of the publishers that, as the series moves into its third decade, it will continue to meet the need and demand for historical writings on the region and that the fruits of the scholarship of new generations of historians will reach a wider reading public.

Other titles in this series are listed at the back of the book.

Democracy Shelved:
Great Britain, China, and Attempts at Constitutional Reform in Hong Kong, 1945–1952

STEVE YUI-SANG TSANG

With a Foreword by David M. MacDougall

HONG KONG OXFORD NEW YORK
OXFORD UNIVERSITY PRESS
1988

Oxford University Press

Oxford   New York   Toronto
Delhi   Bombay   Calcutta   Madras   Karachi
Petaling Jaya   Singapore   Hong Kong   Tokyo
Nairobi   Dar es Salaam   Cape Town
Melbourne   Auckland

and associated companies in
Berlin   Ibadan

OXFORD is a trade mark of Oxford University Press

First published 1988
Published in the United States
by Oxford University Press Inc., New York

British Library Cataloguing in Publication Data

Tsang, Steve Yui-Sang
Democracy shelved: Great Britain, China
and attempts at constitutional reform in
Hong Kong, 1945–1952. — (East Asian historical
monographs).
1. Hong Kong — Politics and government
I. Title   II. Series
354.51'25   JQ675
ISBN 0-19-584175-1

Library of Congress Cataloging-in-Publication Data

Tsang, Steve Yui-Sang, 1959–
Democracy shelved: Great Britain, China, and attempts at
constitutional reform in Hong Kong, 1945–1952 / Steve Yui-Sang Tsang;
with a foreword by David M. MacDougall.
p.   cm.—(East Asian historical monographs)
Bibliography: p.
Includes index.
ISBN 0-19-584175-1: $34.00 (est.)
1. Hong Kong—Politics and government.   2. Representative
government and representation—Hong Kong.   3. Hong Kong—
Constitutional law.   4. Great Britain—Relations—Hong Kong.
5. Hong Kong—Relations—Great Britain.   6. China—Relations—Hong
Kong.   7. Hong Kong—Relations—China.   I. Title.   II. Series.
JQ675.A1T72   1988
951'.25042—dc19   87-34732
CIP

Printed in Hong Kong by Calay Printing Co. Ltd.
Published by Oxford University Press, Warwick House, Hong Kong

For my mother, Chan Kwok-wing, and my friends,
Freddie Madden and Mark Elvin

Foreword

BY the late 1940s, the efforts of the administration which had already laid the foundations of Hong Kong's post-war economic miracle began to be seriously directed towards the delicate question of constitutional change. The reasons why no change in fact resulted are explored in detail in this excellent book. Dr Tsang has thus filled a gap in the colony's history which future students of the period might have found rather puzzling.

Hong Kong enjoyed unique advantages in the immediate post-war years. Her rehabilitation had been rapid and successful, putting her far ahead of her economic competitors and drawing further away all the time. Her industrious citizens were content to leave politics to lie dormant whilst they concentrated on the business of producing the goods needed by the post-war world.

By the beginning of 1947, satisfied that Hong Kong was safely on the way back to ultimate economic success, the colonial administration felt able to give some attention to political matters, in particular the question of introducing an elective form of government in the colony. The call for independence and self-government was beginning to echo around the colonial world, and London had joined in the chorus. However, in Hong Kong, there was a powerful objection: its citizens wanted none of it. Was there, then, a means by which they could be persuaded to accept it, in spite of some obvious dangers — perhaps gradually, by careful degrees?

Dr Tsang's carefully researched book records just how that question was answered by the responsible authorities in London and in the colony itself. It also provides a fascinating insight into the workings of the administrative system developed over many years between London and the overseas territories for which she was responsible. His study of the evidence has led Dr Tsang to the conclusion that the final answer represented a missed opportunity. A stronger attempt might and should have been made to persuade Hong Kong to embark on some measure of elective government by at least the early 1950s, despite the fact that the citizens of Hong Kong were wary of such changes. Gradual reforms — for instance, the election of members to bodies entrusted with municipal duties such as public health and planning — might have been implemented successfully. This was in fact the idea behind Sir Mark Young's original proposal, and had he been able to remain Governor of Hong Kong for a year or two longer, something along those lines most probably would have evolved, in spite of some obvious dangers.

That it did not evolve, Dr Tsang argues, was primarily due to the change in the colony's governorship during the critical period of the late 1940s. It is difficult not to agree with this view. Young had in high degree the qualities of imagination and personality needed to gain public support for his proposed constitutional changes. His successor, though longer acquainted with Hong Kong than Young and a competent enough civil servant, was less inclined to be adventurous or innovative. Perhaps inevitably, Grantham's professional caution was further reinforced by the views of the unofficial members of the Legislative and Executive Councils, who were reluctant to change the foundation of what was proving to be a winning economic hand.

Hong Kong was, in fact, a paradox. Its citizens, who were overwhelmingly Chinese and potential merchants and entrepreneurs to a man, were drawn to the territory by its carefully created and maintained commercial climate. Hong Kong was the one place, for many hundreds of miles around, where a signed contract could be enforced in the courts without fear or favour. The fine harbour and the modern facilities and services aside, the fundamental attraction for what we now call the wealth creators was the integrity and accessibility of the courts. Yet the Chinese residents who used the British courts in this British colony confidently and freely did not really think of themselves as anything other than Chinese. Although few spoke of it, everyone was aware of 1997 and all that that date implied. However, 1997 was almost half a century away and the typical resident was not tempted to become embroiled in politics. Many residents gave generous public service to the community in which they lived and worked, but political activity was best left alone, especially in light of the difficult conditions existing on much of mainland China in the years immediately after 1945.

The difficulty involved in foisting any measure of self-government on a community that did not want it should not be under-estimated. Governor Grantham and his advisers emphasized, too, the dangers of political change in a rapidly changing world. The dangers were certainly real enough; and perhaps Grantham was not wholly wrong in the short-term, since in its substantially unchanged political form, Hong Kong has since enjoyed more than a quarter of a century of unparalleled prosperity. However, this prosperity has been created and sustained, of course, by factors other than the absence of representative government.

The missed opportunity of the late 1940s, as identified by Dr Tsang, did not become fully apparent until the 1984 negotiations which decided the conditions of the termination of the New Territories lease. The citizens of Hong Kong were banned by the Chinese government from participating in these negotiations which directly and vitally affected their future. How different might their position have been if election by secret ballot of members to both the Legislative and Executive Councils (to say nothing of the lesser public bodies) had been in operation and accepted as commonplace since 1950. This is not to say that the outcome of the negotiations might have been different, but at least the people of Hong Kong could have spoken directly to the conference through their elected representatives and could not have been banished from the chamber. At least, not so easily.

The implications of that situation give Dr Tsang's book added importance and topicality — not least, one would have thought, in Taiwan. The book is a valuable contribution to a complicated subject. I admire it greatly.

DAVID M. MACDOUGALL
May 1987

Preface

SINCE the Sino-British Joint Declaration on the future of Hong Kong came into force on 27 May 1985, it has become clear that the political system of Hong Kong will be changed with the transfer of its sovereignty from Great Britain to the People's Republic of China. Chinese officials and their advisers have already begun to examine various issues relating to the constitution of the territory as they prepare to draft the Basic Law for Hong Kong Special Administrative Region, which will come into being on 1 July 1997. However, in 1982, even before the Joint Declaration was agreed, the government of Hong Kong had begun to put into force gradually certain political reforms. The first such reform was the District Administration scheme, which provided, for the first time, direct elections of representatives on a universal franchise at the District or local level. Further changes occurred in the latter part of 1985, when a new system for the indirect election of all unofficial members to the Legislative Council was implemented. The government has also committed itself to reviewing in 1987 the working of the present constitution. These changes have stimulated considerable public interest in and discussion on the subject of political reform.

Readers who are familiar with or interested in the recent changes in the territory will, I am sure, be interested to learn that many of the arguments and even some of the proposals for political reform in Hong Kong being discussed today were debated in the 1940s and 1950s. Although it is not my intention to examine in detail the current debate on constitutional advancement in Hong Kong, I shall highlight very briefly a few issues which have continued since the 1940s to focus the attention of policy makers and others interested in the political development of Hong Kong.

The question which possibly attracts most attention is that of representation of the local inhabitants in the legislature. The most controversial current concern in this connection is whether direct elections should be introduced into the Legislative Council. However, this is not a new problem. In the 1940s, British officials and their advisers studied this issue carefully. In 1949, for example, the unofficial members of the Legislative Council and the Hong Kong Reform Club, amongst others, made separate representations for the introduction of direct elections to the Legislative Council. Although both the Unofficials and the British authorities at that time would only consider enfranchising British subjects, as was then the general practice in the British Empire, the Reform Club went a step further and asked for direct elections for both British subjects and Hong Kong citizens, a category which included non-British long-term residents. For reasons which are explained in the text, none of these proposals were acted upon. At that time, British officials generally saw the Reform Club as being ill-advised; however, with the passage of time, the spirit of the Reform Club's proposal has been accepted. If Hong Kong is to have direct elections, it is no longer acceptable to restrict the vote to British subjects.

A related but somewhat different issue is that of functional representation. Although there was no mention of this term in the 1940s, the idea it entails was examined closely. At first, British and Hong Kong officials in wartime London planned to expand the pre-war practice of indirect elections, which provided the Unofficial Justices of the Peace and the Hong Kong General Chamber of Commerce with the privilege of nominating one of their respective members to sit on the Legislative Council. When Sir Mark Young resumed his governorship of Hong Kong in 1946, he developed the idea further. Young proposed that an elected municipal council be introduced. Amongst his many propositions, he recommended that bodies such as registered trade unions, the Hong Kong Chinese Chamber of Commerce, the Hong Kong General Chamber of Commerce, the University of Hong Kong, the Hong Kong Residents' Association, the Kowloon Residents' Association, and the Unofficial Justices of the Peace be empowered to nominate unofficial members to the proposed municipal council for the representation of the interests of various professional and ethnic groups. In addition, Young suggested that the proposed municipal council also be granted the privilege of nominating two of its unofficial members to the Legislative Council. In other words, he advised that four out of the eight unofficial members of the legislature should be elected indirectly. Even after Young's successor, Sir Alexander Grantham, rejected Young's municipal scheme, Grantham also at one stage proposed that some of the above-mentioned bodies be allowed to indirectly elect six out of the eleven unofficial members to represent various communities in the Legislative Council. Without putting too fine a point on it, and excluding the British officials' concern that a balance of seats be kept amongst the various ethnic communities in Hong Kong's councils in the 1940s and 1950s, the idea of indirect elections scrutinized at that time is in essence not dissimilar to the present-day notion of functional representation.

Another aspect of the current debate on development of the Legislative Council which is reminiscent of previous deliberations is that of maintaining a balance of power between the various categories of members. The current argument is that if there were to be direct elections to the Legislative Council, some local politicians might attempt to secure their election by offering so-called 'free lunches' to the lower income groups, which might scare investors into leaving the territory. Holders of this view believe that it would be necessary to have indirectly elected or even appointed Unofficials in order to counter the influence of some of the 'less responsible' elements and to represent the interests of the investors, which they deem to coincide with those of the community at large. This particular argument is reminiscent of the consideration put forward by Governor Grantham in 1949, when he recommended to the British government that direct elections be introduced to the Legislative Council. Grantham pointed out that the power of the elected and the Chinese unofficial members — the less reliable elements, in his eyes — to defeat the government in the Council should be effectively removed by appointing so many official and nominated unofficial members that the government would almost always carry the day in the Council.

Another current question is whether the Governor should continue to preside over Legislative Council meetings. It has now been suggested that a Speaker should be selected in due course to serve as Chairman of the Council, as the Governor steps down from its presidency. Once again, this suggestion echoes a proposal of the 1940s. When Young's municipal scheme was submitted to Secretary of State Arthur Creech-Jones for approval, Creech-Jones minuted that he preferred to introduce direct elections and a Speaker elected by the Unofficials to the colonial legislature. Like Creech-Jones' many other views on the issue, it was received with scepticism by British officials and was effectively ignored.

Another important subject of the present-day debate is whether Hong Kong should gradually establish a ministerial or quasi-ministerial system of government. This matter, too, has previously been looked into carefully. After Governor Young submitted his municipal scheme to Creech-Jones for approval in late 1946, Creech-Jones minuted that '[s]ome unofficial members from the Legislative Council should be invited to serve on the Government Executive and should be associated with the affairs of particular departments of the Government, the affairs of which they can handle in the Legislative Council, and *gradually* [sic] assume some responsibility in the Executive'. What Creech-Jones advocated was, in essence, a ministerial system of government. This idea was not acted on, partly because of resistance from some senior Colonial Office officials, partly because Creech-Jones was preoccupied with other more pressing issues, and partly because Young forcefully argued that once his proposed municipal council — which he intended in due course to turn into an alternative to the colonial government — had been set up, a ministerial system along Creech-Jones' line would be unnecessary. Young did not oppose the idea of ministerial government as such; he merely deemed it superfluous in the circumstances.

In a small place such as Hong Kong, a question which cannot be avoided is whether there should be several layers or levels of representation and administration. The present system of three tiers of representation — at the District Boards, the Urban and District Councils, and the Legislative Council levels — but only one administrative machinery has drawn some fire on several grounds. It has been argued that Hong Kong has no room or need for three tiers of representation: either the District Boards or the Urban and District Councils should be abolished in order to avoid wastage and overburdening of the general public with frequent elections. An altogether different contention is that, since the central government has maintained a monopolistic control over the administration at all levels, the District Boards as well as the Urban and District Councils are really sham, rather than genuine, channels of representation. The Boards and the Councils are considered to be ineffective because they are in practice merely government agencies with consultative roles.

Holders of this view believe that these bodies should be replaced by municipal and District Councils which have their own independent sources of finance and administrative arms. This concern over the overlapping of various councils is reminiscent of the discussions of the 1940s. At that time, some officials doubted whether there was room in the territory for a fully fledged municipal council. A committee of inquiry appointed by the British government also suggested that the Legislative and Urban Councils be merged to form a State Council, which would have both legislative and executive authorities. Others, like Governor Young, took a different view. As he saw it, given the very delicate political situation prevailing in Hong Kong, it was advisable to defer the establishment of a representative government at the central level until a similar experiment at the 'municipal' level had been successfully carried out.

The delicate political situation which Young (and his colleagues) referred to was the result of two factors: the existence of a time limit to British jurisdiction over the leased territory of Kowloon (now known simply as the New Territories), and the possibility of the supporters of the two major Chinese political parties — the Kuomintang and the Chinese Communist Party — introducing Chinese national politics into the directly elected councils of Hong Kong. The nature of the problems created by these factors has altered considerably over the years, but they continue to restrain the hands of British officials. In the past, British officials were wary of reforming the Hong Kong government, often because they were worried that such actions might send the wrong message to China with regard to the political status of the territory, thereby provoking unwanted countermeasures. Since the signing of the Joint Declaration, the new primary concern of British officials must be the ensuring of a smooth transition of government in 1997. As far as subversive activities of the two Chinese parties and their supporters are concerned, the danger of Chinese Communist infiltration has become less and less relevant. (The concern of the Hong Kong people about the roles which agents and supporters of the People's Republic will play in the near future and after 1997 is a completely different matter.) Nevertheless, however correctly Kuomintang supporters in Hong Kong might have behaved in the recent past, policy makers in Hong Kong and London cannot ignore completely the possibility of some Kuomintang die-hards stirring up trouble during local elections or in the community at large prior to 1997. This delicate political situation of the territory will no doubt continue to affect the direction of reform in the foreseeable future.

Despite the similarities mentioned above between the current discussions of reform and those of the 1940s, there are also of course many differences which, for various reasons, will not be examined here. I may perhaps emphasize that this book is above all a study of events past. Valuable lessons about reforming the government in Hong Kong can and, indeed, should be learned by the present generation, but readers will undoubtedly concede that the circumstances in Hong Kong have changed markedly in the last forty years.

To begin with, the nature of the uncertainty about Hong Kong's future has changed. In the 1940s, the uncertainty was as to whether Britain would hand over Hong Kong to the Chinese government under the Kuomintang. This question was temporarily settled on the eve of the Communist take-over in China when the British Cabinet decided to employ the recently reinforced Hong Kong garrison of 30,000 troops supported by tanks, artillery, and powerful air and naval units to back up its Hong Kong policy of August 1949. The gist of this policy was that whilst His Majesty's government was ready and prepared to discuss Hong Kong's future with a Chinese government that was friendly, united, and stable, it deemed that such a government did not exist in China at that time and it did not expect it to appear in the foreseeable future. Since the signing of the Joint Declaration on Hong Kong's future in 1984, however, it has become clear that Hong Kong will be returned to China in 1997. The uncertainty which now exists relates to how much Hong Kong after 1997 will resemble Hong Kong of today.

Furthermore, in the last forty years, the character of Hong Kong's inhabitants has undergone changes akin to the biological process of metamorphosis. Three or four decades ago, the local inhabitants were mostly illiterate or semi-illiterate immigrants, who had for the most part come to Hong Kong either as Chinese sojourners before 1949 or subsequently as political or economic refugees. They were not to any extent noticeably different from their fellow countrymen elsewhere in China. Today, the vast majority of Hong Kong's population are educated, locally born permanent residents who are easily distinguishable from the

Chinese on the mainland. The spread of literacy and education has enabled the local people to participate intelligently in local elections and public affairs. More importantly, whilst the over-whelming majority of those locally born residents of Chinese origin did not claim British nationality during the immediate post-war years, the majority have now done so. (The fact that, according to British law and whether or not these locally born residents claim British nationality, they are British nationals is beside the point.) Indeed, although most of Hong Kong's present-day citizens of Chinese extraction also consider themselves to be Chinese, they carefully differentiate between themselves — as Hong Kong Chinese or 'HongKongers' — and the inhabitants living north of the border — who are 'mainland Chinese' or, simply, 'the mainlanders'. This emergence of a Hong Kong identity, which became increasingly discernible in the 1970s and is unmistakable in the 1980s, is particularly pronounced amongst the younger and better educated. They do not think of themselves as Chinese sojourners in, but as citizens of, Hong Kong. It is also becoming less and less correct to regard the Hong Kong people as apathetic with respect to politics. Hong Kong is finally developing a political culture of its own. The rapid rise of pressure groups and quasi-political organizations in the last fifteen–twenty years is perhaps the best testimony to this change. They have been clamouring for greater accountability of the administration for years, and are now increasingly asking for a share in the government itself. This rapid transformation strained Hong Kong's political system to the limit before the recent re-forms were introduced to meet some of these rising public aspira-tions. Whilst it is still too early to make any conclusive judgement on the effectiveness of these changes, it must be emphasized that since the flood gates for democratic reform have been opened, they cannot now be shut.

A related but different aspect of the changed circumstances is the result of Hong Kong's spectacular successes in economic and social developments during the same period. Sir Cecil Clementi, Governor in the late 1920s, aptly summarized the crucial consid-eration regarding local government in pre-war Hong Kong: he said that the colony was so small and compact that it was in effect a large township, and were he and his principal assistants to divest themselves of municipal duties, there would be little left for them to do. Although Hong Kong was no longer at the end of the Second World War a large township, a municipal government was still arguably adequate for administering matters of purely local concern, leaving the imperial government in London to take care of external affairs. However, the tremendous growth in Hong Kong's population, the increasing need for the territory to look after its economic and other relations with the outside world, the gathering pace of urbanization, the transformation of Hong Kong into a modern metropolis — a thriving commercial, industrial, and financial centre — have rendered Clementi's once astute observation wholly inappropriate. Whilst the physical size of the territory has not expanded to any extent, the requirements of modern Hong Kong have made it right and proper that the best means of sharing the burden of administration between the Hong Kong government and local authorities begin to be examined. The availability of wealth as a result of economic development has also provided ample resources for the territory to adopt a more democratic (which is often also more expensive) system of government, if that is deemed to be desirable.

Another important change which has taken place concerns the nature of the activities of the Kuomintang and the Chinese Communist Party in Hong Kong. In an important sense, these two parties have exchanged the positions they once held in the territory. Before 1949, the Kuomintang was the ruling party of China and in Hong Kong it was seen by some local Chinese inhabitants as an alternative to the government of Hong Kong. This attitude gradually changed as the Kuomintang lost the 'mandate of heaven' on the Chinese mainland in 1949 and retired to Taiwan. Although the local inhabitants continued to be wary of the Chinese Communists after 1949, this attitude began to change quite markedly following the visit of President Richard Nixon of the United States to Peking (Beijing) and the admittance of the People's Republic of China to the United Nations in the early 1970s. By way of contrast, in present-day Hong Kong, most if not all of those local residents who conceive of an alternative to the colonial government think not of the Kuomintang regime in Taiwan but of the Chinese Communist government in Peking.

Also of importance is the attitude of the Chinese government to the question of political development in Hong Kong. Since the signing of the Sino-British Joint Declaration, it has been generally accepted that any proposals for reforming the constitution of the territory before 1997 should be compatible as far as possible with the Basic Laws for Hong Kong Special Administrative Region conceived by the Chinese government. In other words, while the Hong Kong government must not be allowed to become a lame duck, the British government must also take into account the attitude of the Chinese government on the subject, as expressed either through normal diplomatic channels or through the Sino-British Joint Liaison Group on Hong Kong.

In view of the changes highlighted above, to what extent are the arguments and considerations which attracted the attention of policy makers and others interested in the subject in the 1940s and 1950s still valid today? The answer to this question is, I believe, best left to individual readers. Nevertheless, this study of, as it were, a non-event remains instructive.

STEVE TSANG

Contents

Plates

Acknowledgements

IN undertaking the research for this book, I have received generous assistance from various individuals and institutions. I am most grateful to Dr A. F. Madden (my supervisor whilst I was a student in Oxford) and Professor K. E. Robinson for providing much invaluable help.

I am indebted to Mr Brook Bernacchi, Mr C. B. Burgess, Sir Sydney Caine, Sir Jack Cater, Mr Percy Chen, Sir John Cowperthwaite, Dr G. A. C. Herklots, Mr Leung To, Mr David MacDougall, Mr Ng Wah, Lady Paskin, Mr Patrick Sedgwick, Mr J. B. Sidebotham, and Reverend Alastair Todd for kindly speaking of their own experiences. Mr MacDougall in particular has most kindly written a Foreword for the book, granted me full access to his personal collections of photographs and papers relating to Hong Kong, and given me permission to reproduce some of his photographs as illustrations. I am also obliged to Dr G. W. Catron for use of his valuable thesis on Hong Kong and China; to Professor D. W. Y. Kwok for reproducing a letter of his father, Mr T. W. Kwok; to Mrs J. Snowdon for giving permission to quote from the private papers of her father, Sir Franklin Gimson; to Sir Adrian Swire for access to the diary of his father, Mr John Swire; and to the *South China Morning Post* for permission to reproduce in the book selected photographs from the *Post*, the *Hong Kong Telegraph*, and the *China Mail*.

I am also grateful to the keepers and staff of various institutions wherein I collected most of my material and some of my thoughts. I must especially thank those connected with the Rhodes House Library (Oxford), the Public Records Office (Kew), the Public Records Office of Hong Kong, the School of Oriental and African Studies (London), Hung On-to Memorial Library (Hong Kong), the University Services Centre (Hong Kong), and the Fung Ping Shan Library (Hong Kong).

I am grateful in a different way to Mr S. R. Bailey, Mr Alan Bell, Dr Alan Birch, Mrs Sandra Lee-Birch, Miss Eileen Bryan, Miss Jane Chan, Miss Veronica Chang, Dr W. E. Cheong, Miss Mathalina Cheung, Mr Ian Diamond, Dr Mark Elvin, Mr K. K. Li, Dr Alfred Lin, Mr Cyril Lin, Dr Thomas Lau, Mr T. L. Lui,

Mrs R. H. McLean, the staff of Oxford University Press, Professor R. E. Robinson, Dr James Tang, Dr Herman Tsui, Mr Peter Yeung, and Professor L. K. Young for providing various kinds of assistance at different stages of my work.

Finally, I am very thankful to John Swire and Sons Limited and Cathay Pacific Airways for providing generous financial support in the first three years of my research; to the administrators of the Beit Fund for providing financial backing for my fourth year of work; and to the Management Committee of the Chinese Studies Centre of Wolfson and St Antony's Colleges for further support which enabled me to complete my work.

Abbreviations

CAB	Cabinet papers
CAPMC	China Association Papers — Minutes and Circulars
CATC	Central Air Transport Corporation
CNAC	Chinese National Aviation Corporation
CO	Colonial Office
CPIS	*Colonial Political Intelligence Summary*
CPR	*Chinese Press Review*
DEFE	Chiefs of Staff Committee Papers
FEER	*Far Eastern Economic Review*
FE(O)C	Far Eastern (Official) Committee
FO	Foreign Office
FRUS	*Foreign Relations of the United States*
HC DEB	*House of Commons Debate*
HKCGCC	Hong Kong Chinese General Chamber of Commerce
HKFTU	Hong Kong Federation of Trade Unions
HKGCC	Hong Kong General Chamber of Commerce
HKPU	Hong Kong Planning Unit
HKRS	Hong Kong Record Series
HKSH	*Hong Kong Sunday Herald*
HKT	*Hong Kong Telegraph*
HKTUC	Hong Kong Trade Union Council
HSP	*Hwa shiang pao (Chinese Merchant's Daily)*
HSWP	*Hsin sheng wan pao (New Life Evening Daily)*
JCPS	*Journal of Commonwealth Political Studies*
JHKRAS	*Journal of the Hong Kong Branch of the Royal Asiatic Society*
JICH	*Journal of Imperial and Commonwealth History*
JOS	*Journal of Oriental Studies*
JRCAS	*Journal of the Royal Central Asian Society*
KMJP	*Kuo min jih pao (The National Times)*
KSJP	*Kung shang jih pao (Industry and Commerce Daily)*
NRM	*Hsin-wen-yen-chiu tsu-liao tsung-ti-shih-erh-chi (Newspaper Research Materials*, Vol. 12)

PRO	Public Records Office
SCMP	*South China Morning Post*
STC	Smaller Territories Committee
STJP	*Sing tao jih pao* (*Star Island Daily*)
TKP	*Ta kung pao* (*The Impartial Daily*)
WHP	*Wen hui pao* (*The Literary Daily*)
WKYP	*Wah kiu yat pao* (*Overseas Chinese Daily News*)

Chronology

1943

August The Hong Kong Planning Unit (HKPU) is formed in London under N. L. Smith, a retired Colonial Secretary of Hong Kong.

1944

September D. M. MacDougall (a Hong Kong Cadet) becomes Head of HKPU.

1945

May The Colonial Office, HKPU, and the China Association begin examining seriously the proposals for post-war constitutional changes in Hong Kong.

July G. H. Hall (Labour) takes over from Oliver Stanley (Tory) as Secretary of State for the Colonies after the general election.

August Japan surrenders.

F. C. Gimson, Colonial Secretary of Hong Kong during the Second World War, proclaims himself Officer Administering the Government in Hong Kong.

September British Military Administration under Rear Admiral C. H. J. Harcourt begins.

MacDougall (Chief Civil Affairs Officer) and his team from the HKPU gradually take over all civil responsibilities from Gimson's interim government.

The local Japanese Commanders formally surrender to Harcourt at Government House.

October–November Units of the Chinese Army (8th and 13th Army Corps) pass through Hong Kong to north China without any major incident.

November The Chinese Special Commissioner for Hong Kong, T. W. Kwok, arrives and assumes his duties.

Hong Kong reopens to normal trading.

1946

January   Sir Mark Young (Governor since 1941) begins participating in the Colonial Office deliberations concerning constitutional changes in Hong Kong.

May   The British Military Administration ends and Harcourt leaves Hong Kong.

Young returns as Governor, restores civil government, and announces His Majesty's Government's intention to give the local inhabitants a greater say in the management of their own affairs.

T. M. Hazlerigg is appointed Special Adviser to the Governor and a member of the Executive Council to assist in matters relating to constitutional reform.

MacDougall becomes the first post-war Colonial Secretary.

August   Young speaks to the people of Hong Kong over the radio, outlining his preliminary proposals for reform.

October   Arthur Creech-Jones (Labour) succeeds Hall as Secretary of State.

Young submits his recommendations for reforming the government of Hong Kong (the Young Plan) to the Secretary of State.

1947

May   Young retires from the governorship of Hong Kong. Hazlerigg has retired earlier.

July   The Hong Kong and Macau Branch of the Kuomintang holds its first-ever annual general meeting in Hong Kong.

His Majesty's Government publicly approves in principle the Young Plan.

Sir Alexander Grantham arrives as the new Governor of Hong Kong.

October   J. H. B. Lee is appointed Assistant Colonial Secretary (Special Duties) for work in connection with constitutional reform.

November   The Hong Kong government sends three draft bills required for the implementation of the Young Plan to the Colonial Office.

December The Kowloon Walled City Incident begins.

1948

January The Shameen Incident occurs.

The Kowloon Walled City Incident ends.

February Lee is appointed Acting Post Master General, relinquishing his appointment as Assistant Colonial Secretary (Special Duties).

March Lord Listowel, Minister of State for Colonial Affairs, visits Hong Kong.

Grantham addresses the Legislative Council, saying that everyone had been guilty of underestimating the practical difficulties in implementing the Young Plan.

April The Treasury control over Hong Kong ends.

November W. J. Carrie is appointed Special Adviser responsible for constitutional development but without a seat at the Executive Council.

1949

January The Hong Kong Reform Club is formed.

April The *Amethyst* Incident begins.

May His Majesty's Government announces its decision to reinforce the defence forces of Hong Kong.

The Hong Kong Chinese Reform Association is formed.

MacDougall leaves Hong Kong for retirement.

June The *Hong Kong Government Gazette* publishes the three draft bills concerning the implementation of the Young Plan.

A. V. Alexander, Minister of Defence, visits Hong Kong.

The unofficial members of the Legislative Council propose to substitute for the Young Plan the reform of the Legislative Council (the Unofficials' alternative).

The Reform Club petitions the Governor for, amongst other things, a directly elected Legislative Council.

July   The Hong Kong Chinese Reform Association, the Chinese Manufacturers' Union, the Kowloon Chamber of Commerce, the Kowloon Chinese Chamber of Commerce, and 138 other Chinese organizations petition the Governor for constitutional changes at both the central and municipal levels.

August   The *Amethyst* Incident ends.

Grantham reports to the Secretary of State, recommending the substitution of Young's proposals by the Unofficials' alternative.

September   Mao Tse-tung proclaims the establishment of the People's Republic of China in Peking.

October   Chinese Communists take over Canton. Irregular Chinese Communist forces reach Hong Kong's border.

The Smaller Territories Committee is formed.

December   The Tramway Strike begins.

1950

January   The United Kingdom recognizes the People's Republic of China.

T. W. Kwok's office closes.

The Russell Street Incident occurs.

February   The Tramway Strike ends.

James Griffiths (Labour) succeeds Creech-Jones as Secretary of State after the British general election.

May   Hong Kong introduces immigration control against Chinese entering from mainland China.

W. J. Carrie is appointed Acting Chairman of the Urban Council until 21 June, presumably in addition to his appointment as Special Adviser for constitutional reform.

June   Grantham arrives in the United Kingdom on leave and begins a series of meetings with British ministers and officials about constitutional changes in Hong Kong.

The Korean War begins.

October   China enters the Korean War.

Grantham returns to Hong Kong, having reached an understanding with the Colonial Office on the substitution of the Unofficials' alternative by another scheme (Grantham's 1950 proposals).

December The United States imposes a virtual embargo on selected exports to Hong Kong.

The Foreign Office asks the Colonial Office not to implement Grantham's 1950 proposals at this stage.

1951

February Grantham agrees with the Foreign Office's suggestion that the reform proposals for Hong Kong be put aside.

China imposes a system of entry and exit control to Kwangtung province, easing the pressure of population movement into Hong Kong.

October Oliver Lyttelton (Conservative) takes over as Secretary of State after Labour's defeat in the British general election.

December Lyttelton visits Hong Kong and discusses matters concerning political reform with various local organizations.

1952

February Grantham requests permission from the Colonial Office to resume preparations for constitutional reform.

March Riots along Nathan Road occur.

May The British Cabinet approves Grantham's 1950 proposals for reform.

The first post-war Urban Council elections take place.

June Grantham sends a telegram to the Colonial Office to express the Unofficial Councillors' 'fear' of constitutional reform.

July Grantham returns to England on leave and requests Lyttelton to persuade the Cabinet to withdraw its earlier approval of reform.

September The Cabinet agrees to drop all major reform for Hong Kong.

October Minor disturbances take place on the national days of the People's Republic of China (1 October) and the Republic of China (10 October).

Grantham returns to Hong Kong.

Lyttelton announces in the House of Commons that reform in Hong Kong would be limited to the Urban Council, any major reform being 'inopportune'. Grantham adds in the Legislative Council that minor reform proposals would be received with sympathy.

1 Introduction

Historical Background

British possession of Hong Kong formally began when Commodore Sir Gordon Bremer of the Royal Navy, acting under the instructions of Captain Charles Elliot (the British Plenipotentiary in China and Chief Superintendent of China Trade), occupied the island on 26 January 1841, in the course of the First Anglo-Chinese War (1840–3). Subsequently, Elliot proceeded to make arrangements for its administration. However, Hong Kong did not become a British colony until after the Treaty of Nanking was ratified, almost two and half years later on 26 June 1843.

When Britain again defeated China in the Second Anglo-Chinese War (1856–60), she also acquired — in the first Convention of Peking (1860) — Kowloon peninsula. The southern tip of Hsin An county on the Chinese mainland, Kowloon is across the harbour from Victoria (now generally known as Central District). Hong Kong's territorial jurisdiction expanded further in 1898 after China's defeat in the Sino-Japanese War (1894–5), which had triggered off the so-called scramble for concessions in China. In order to better defend the colony, Great Britain leased for ninety-nine years from China — in the second Convention of Peking (1898) — that part of Hsin An county north of Kowloon and south of the Sham Chun river (including 235 adjacent islands). The lease for this territory (known as the New Territories), which constitutes ninety per cent of the colony's total area, expires on 30 June 1997.

The island of Hong Kong and its dependencies were established as a British colony, with a Crown colony system of government, by the Hong Kong Charter of 5 April 1843, which came into effect when Sir Henry Pottinger succeeded Elliot as Governor of Hong Kong on 26 June. The Royal Instructions to Pottinger, dated 6 April of the same year, provided detailed guidance as to the working of the constitutional arrangements set out in the Charter. These two documents made up the constitution of Hong Kong. Governor Pottinger was to constitute and preside over a Legislative and an Executive Council by nominating three officials for appointment to each of the Councils. The essential features of the constitution were that there would be no unofficial representation in the Councils, and that the Councils would be able to discuss only such matters as the Governor chose to place before them. The Governor could, therefore (at least in theory), act at his absolute discretion as long as he enjoyed the support or confidence of the Secretary of State, to whom he was directly answerable.[1]

This constitution gradually evolved, while the colony developed during the next hundred years from a small fishing harbour and haven for pirates at the time of the British occupation into an important entrepôt and a modern city. The first major attempt to introduce political reform was made shortly after Hong Kong was founded as a colony. When Sir John Davies (Governor, 1844–8) tried in 1845 to raise revenue by, amongst other things, levying rates on property for policing and other municipal services, British merchants in the colony presented a memorial to the Secretary of State, Lord Stanley, calling for some undefined form of municipal self-government. The British merchants did not refer in their memorial to electoral arrangements or the representation of the Chinese community, and they were not particularly interested in taking up municipal responsibilities; their object was to resist Davies' proposed financial imposition. In this matter, Davies enjoyed the support of both the Secretary of State and Sir James Stephen, the Permanent Under-Secretary at the Colonial Office. However, the 1847 Select Committee of the House of Commons on China Trade, which was sympathetic to the views of the merchants, also interested itself in the issue. It recommended the introduction of some system of municipal government for the British residents of Hong Kong. This recommendation encouraged the British mercantile community to renew its demands in 1849 in a petition to Parliament. Sir George Bonham, who had succeeded Davies as Governor in 1848, and Earl Grey, the new Secretary of State, were ready to establish a municipal council in Hong Kong provided that the inhabitants were willing to bear the full cost of policing. This the British community refused to do, and they did not renew their demands after March 1851.[2]

Partly as a result of these local agitations, two Unofficials were admitted to the Legislative Council in 1850. This was followed in 1884 by the admission of the first Chinese Unofficial to that

Council on a permanent basis,[3] and the formal granting to the Hong Kong General Chamber of Commerce and the Justices of the Peace of the privilege to submit to the Governor the name of one of their members for the purpose of his nomination to the Legislative Council. By this time, the Legislative Council had expanded to include six official members (excluding the Governor) and five unofficial members; the membership of the Executive Council had also been increased to six officials.

Further major reform occurred in 1896, partly as a result of agitation by the British mercantile community for greater representation. In 1894, T. H. Whitehead, an unofficial member of the Legislative Council, organized a petition of 363 ratepayers to Lord Ripon, the Secretary of State, asking for the free election of representatives of British nationality to the Legislative Council and a majority of such members in the same Council. In an important sense, this attempt resembled the earlier one, the main objective being to prevent further increases in taxation and government expenditure, which had risen gradually during the previous decade. Despite their reliance upon liberal arguments, the petitioners wanted in effect (as Endacott expressed it) 'an oligarchy of some 800 voters' who were mostly, if not exclusively, expatriate Britons. There was no question of representation of the Chinese community, who numbered 211,000 out of the total population of 221,400, and who contributed most of the rates collected in the colony. The petitioners' proposals received no support from the Governor, Sir William Robinson, from Lord Ripon, or from his successor as Secretary of State, Joseph Chamberlain. Robinson, Ripon, and Chamberlain shared the view that Hong Kong's status as an imperial station on the border of a foreign country rendered it unsuitable for any form of self-government. They also regarded it as invidious and inequitable that the Chinese community should be left to the mercy of an oligarchy of British traders. The outcome was that although the petition was rejected, two non-official expatriate Britons were appointed in 1896 to the Executive Council and an additional Chinese was admitted to the Legislative Council.[4]

Another attempt by the expatriate British residents to increase their representation in the government was made during the First World War. It, too, was led by an unofficial member of the Legislative Council, H. E. Pollock, whose appointment to the Council at that time was based on the recommendations of the

Unofficial Justices of the Peace. After failing in 1915 to obtain a seat on the Executive Council, Pollock organized a petition to the Secretary of State in January 1916 calling for the appointment of two additional unofficial members to the Executive Council, one of whom was to be nominated by the Unofficial Justices of the Peace and the other by the Hong Kong General Chamber of Commerce. The petitioners also asked for the introduction of an Unofficial majority in the Legislative Council by increasing the number of representatives of the British community from four to eight. The petition was rejected by both the Governor, Sir Henry May, and the Secretary of State, Bonar Law, on virtually the same grounds that had induced their predecessors to turn down the petition of 1894.

Although the demands of the British community for political power for themselves were unsuccessful, the community continued to agitate, resulting in the formation of the Constitutional Reform Association of Hong Kong in May 1917 and the Kowloon Residents' Association in January 1921. Successive Governors and Secretaries of State continued to share the opinions of their predecessors, however, and the agitations finally died down in 1921. From that time, the British mercantile community made no further serious attempts to amend the constitution of the colony.[5] However, the re-examination of Hong Kong's constitution during the First World War led to the original constitutional instruments being superseded, on 14 February 1917, by a new set of Letters Patent and Royal Instructions in order to keep the constitution up to date.[6]

Subsequent changes in the colony's constitutional arrangements had the effect of increasing the representation of the Chinese community. In 1926, Sir Shouson Chow was admitted as the first Chinese unofficial member of the Executive Council. Chow's appointment increased to three the number of unofficial members in the Council.

In all of their attempts before the Second World War to reform the constitution of Hong Kong, the expatriate British merchants and residents of the colony aimed at enhancing their influence over the government. There was no question of 'one man, one vote'. The reformers also generally ignored the representation and interests of the Chinese inhabitants, who constituted the overwhelming majority of the population and taxpayers. This neglect of the Chinese community was one of the main reasons why the imperial government in London and the colonial officials in Hong Kong rejected the demands of the expatriate Britons. However, the British authorities did not have the interests of both the British and the Chinese communities equally at heart. Despite their insistence that the Chinese community not be left to the mercy of an oligarchy of British traders, the interests of the British community were of paramount importance to the authorities. This attitude is hardly surprising since the colony was founded, in W. E. Gladstone's words, 'solely and exclusively with a view to [primarily British] commercial interests'.[7] Although Gladstone somewhat overstated his case, there is some truth in his statement. The Chinese community made no real attempt to obtain representation in the government commensurate with its contribution to and importance in the colony. The sympathies of most of the Chinese residents of Hong Kong at that time, with the exception of a handful who were thoroughly Anglicized, still lay with China.

By the eve of the Second World War, the constitution of Hong Kong provided for an Executive Council of seven official members (including the Governor) and three unofficial members, and a Legislative Council of ten official members (including the Governor) and eight unofficial members. Of the unofficial members of the Executive Council, one was Chinese and two were European. Of the unofficial members of the Legislative Council, three were Chinese, one was a local Portuguese, and three were European. All of the unofficial members continued to be appointed, and two of the three non-Chinese members of the Legislative Council continued to be appointed on the basis of nominations by the Unofficial Justices of the Peace and the Hong Kong General Chamber of Commerce. In 1936, an Urban Council was also set up, consisting of five official and eight unofficial members. Of the unofficial members, two were elected by an electorate drawn up from the jury list (in essence, people literate in the English language); the remaining six seats were divided equally between the Chinese and non-Chinese nominees of the Governor. The Council, chaired by a Cadet (or Administrative Officer) directly responsible to the Governor, was almost entirely responsible for sanitary matters. In spite of its title, the Urban Council was not a municipal council; rather, it was a hybrid of a government department and a large advisory board.[8]

The fundamental political problem in pre-war Hong Kong was

the maintenance of its colonial identity and the satisfaction of the demands of the British mercantile community, without compromising the predominantly Chinese character of Hong Kong. The failure to introduce local government at the municipal level is best explained by Sir Cecil Clementi (Governor, 1925–30):

This Colony is so small and so compact that it is in effect a large township, and the Government of Hong Kong is and must always be mainly concerned with municipal affairs. I regard myself as being in effect mayor of Hong Kong; and were I and the principal officers of this Government to divest ourselves of our municipal duties there would be little left for us to do.[9]

The rise of modern Chinese nationalism following the May Fourth Movement of the late 1910s and the accession to power in China of the Kuomintang in 1927 further complicated the reform of Hong Kong's constitution. Although no Chinese government had ever accepted the loss of Hong Kong to Britain, the nationalist government under the Kuomintang advocated its return to China. Whatever the Kuomintang government's policy with regard to Hong Kong, however, it made no official attempt before the Second World War to secure its return. It was not until 1942, following the loss of Hong Kong to the Japanese, that the Chinese government formally made such a request. Although the British government declined to return the territory, Chinese irredentism over Hong Kong henceforth became an important consideration for the British authorities dealing with constitutional arrangements for the colony.

Three Interpretations of the Attempts at Reform (1945–52)

Whilst original research and critical studies relating to the attempts at reform in Hong Kong immediately after the war are few, there are three main interpretations of the British government's decision in 1952 to abandon, in spite of seven years of planning and preparation, all attempts at major constitutional reform.

The most important pioneering work was done by George Endacott. In *Government and People in Hong Kong: 1841–1962*, Endacott argued that constitutional advancement was adopted as a policy for Hong Kong partly because it was an international obligation under the Charter of the United Nations (Chapter XI, Article 73) and partly because of a genuine desire on the part of the British government to encourage its colonies to share in a partnership of free nations within the Commonwealth.[10] The long and winding course of these attempts at constitutional advancement was seen to be the result of the complexity of the issues involved and the reluctance on the part of Whitehall and Government House to impose reform on the colony against its wishes.[11] Endacott accepted the wisdom of the British government's decision in 1952 to limit changes to the local Urban Council on the grounds that the time was inopportune for any major constitutional change.[12] He stressed that there was no change in the policy of the British government and that the explanation for this turn of events must be sought 'in the general situation in the Far East, which had undergone a radical transformation in the six years between 1946, when the promise was given ... and 1952, when the decision against major changes was announced'.[13]

In an important sense, Endacott failed to appreciate fully the intricacies involved and he consequently provided a rather superficial account of the attempts at reform. He also failed to take into adequate consideration the attitudes of the two Governors involved; the attitudes of local leaders and the general public; the concern of the policy makers to guard against Kuomintang and Chinese Communist infiltration; and the general socio-economic situation within the colony. Endacott did not have access to the official archives and he was reliant upon English-language sources. He was also apparently unaware of the *Chinese Press Review* (prepared and published since 1947 by the American Consulate-General in Hong Kong), which contains translations and/or summaries of selected editorials and news items in the leading Hong Kong Chinese-language newspapers.

C. B. Burgess, a former Colonial Secretary of Hong Kong, provides a more sophisticated interpretation of the attempts at reform in a private paper deposited at Rhodes House Library, Oxford. Burgess traces the reform proposals from their origins to the Second World War. He sees the reaffirmation of 'self-determination' to be a guiding principle for colonial policy in the same way that the production of the Beveridge Report was for domestic policy. The wartime British planners for post-war Hong Kong (then working in London, and including both officials and non-officials) accepted without question His Majesty's Government's overall colonial policy. Their concern for Hong Kong and its people led to the various proposals to introduce constitutional reform in the colony. However, there were no crowds clamouring for the franchise and there were no demagogues agitating for independence. The ordinary Chinese were apathetic, and even the unofficial members of the Executive and Legislative Councils and the business community were sceptical. The activities of the Kuomintang inside the colony, which were seen as liable to turn the gold of reform into the dross of subversion and revolution, caused apprehension in official quarters and contributed to the apparently vacillating manner in which the matter of reform was handled. According to Burgess, it soon became apparent that the real death-knell of reform was neither the relative apathy of the people nor the apprehension of the government, but the simple fact that the government of China would not allow any such constitutional development to take place.

Several of Burgess' colleagues, who had served in the government of Hong Kong during this period, expressed similar views when interviewed for this study. Although this interpretation is now generally accepted for a later period, the question remains whether it is a valid interpretation of the situation in Hong Kong in the late 1940s and early 1950s.

In their book, *Political Reform in Hong Kong — The Past and the Future* (in Chinese), H. T. Cheung and C. K. Lo take the view that the attempt at reform in Hong Kong was the logical consequence of 'the Colonial Office's decision in 1943 to organize self-government in British colonies after the War'.[14] They dismiss political apathy as the crucial factor in the eventual abandonment of all the major reform proposals because of its inconsistency with Britain's commitment to decolonization.[15] They also expound the view that the unofficial members of the Legislative Council, as the representatives of vested interests, rejected in 1949 the existing proposals to establish a full municipal council in Hong Kong and to introduce an Unofficial majority of one in the Legislative Council. According to Cheung and Lo, the Unofficials made a counter-proposal which caused the British authorities to 'hesitate': they suggested that the Legislative Council be changed before any proposal pertaining to a municipal council be examined and, in contrast to the proposals then under consideration, they proposed that the franchise be restricted to British subjects. Cheung and Lo are not very clear and coherent in their interpretation. They appear to suggest that both elements of the Unofficials' counter-proposal were unacceptable to the British government: that is, reform of the Legislative Council, because this might affect the authority of the colonial government, which would in turn alter Hong Kong's position in British Far Eastern strategy; and the restriction of the franchise to British subjects, because the British government had previously rejected all similar proposals.[16]

Cheung and Lo's interpretation is not argued convincingly. The paucity of original research and the book's loose structure also detract from its value. The most intriguing proposition they make is that the unofficial members of the Legislative Council had in effect attempted to sabotage the reform proposals. However, no convincing evidence is produced to support this contention. Moreover, their presentation of the Unofficials' approach makes it more like a 'conspiracy theory' than a conclusion drawn from solid research.

Notes on Sources, the Approach, and References

This book is primarily a case study of British colonial policy in Hong Kong and its constitutional development within the context of the post-war Far East. In principle, at any rate, the study should therefore be based largely on the archives of both the British and the Hong Kong governments. However, while the British archives have been used extensively, most of the Hong Kong government files which deal specifically with this subject are inaccessible. In spite of the kind assistance of Ian Diamond, former Archivist of the Hong Kong Public Records Office, all attempts to obtain access to those documents have been unsuccessful. Only three unimportant Hong Kong government files on constitutional reform are available. One of these files deals with the views of the Commissioner of Police as to whether Young's proposed municipal council should be responsible for the registration of vehicles, and related matters. A second file examines whether the Hong Kong Council of Women should be made a nominating body for the proposed municipality. The remaining file is concerned with the provision of ballot boxes.[17] The 1949 and 1951 Colonial Office files on constitutional reform for Hong Kong are also unavailable. However, British policies for Hong Kong were in most cases formulated after consultation with other government departments in Whitehall, and reference has been made to the archives of those departments, most notably the Foreign Office. Indeed, a significant number of important documents — particularly those dealing with such problems as security and Chinese political activities in the colony — and including those originating in both the Hong Kong government and the Colonial Office, are found only in the Foreign Office (and, occasionally, the Cabinet Office) archives.

Britain's policy on constitutional reform in Hong Kong, while formulated by policy makers in Government House and Whitehall, reflected more than the preferences of the policy makers involved. Events in Hong Kong itself, in neighbouring China, and in the Far East as a whole were important factors which had to be taken into consideration before the direction and content of the proposed Hong Kong reforms could be agreed. The most important considerations were the opinions of the unofficial members of the Executive and Legislative Councils as well as those of the local people; the need first to rehabilitate and then to develop the economy; the activities of the potentially disruptive Kuomintang and Communist Party elements in the colony; the coming to power of a Communist government in China; the change in Anglo-Chinese relations; the outbreak of the Korean War; and the imposition by the United States of severe restrictions on trade with Hong Kong as a result of the Korean conflict. It is, of course, essential that the official British view and understanding of the situation in Hong Kong is not confused with the actual situation. Local newspapers and other publications in both the Chinese and English languages, interviews with people who took part in various events, private papers, and published American documents have therefore been used widely in this study to supplement and verify the interpretations of events in the British (and, to a much lesser extent, Hong Kong) documents.

Public opinion in Hong Kong in the 1940s and 1950s with regard to the question of constitutional reform — or, indeed, to any issue — is difficult to gauge because the vast majority of the population did not express their views at all. However, opinions were sometimes expressed in the form of newspaper editorials, public statements released by various local organizations, letters to newspaper editors, and public petitions to the Governor, and it is possible therefore to assess what might be called 'articulate opinion' of the plans for reform. For the purposes of this study, 'articulate opinion' will be used whenever reference is made to such opinions.

In order to minimize the number of footnotes, references to documents covering the same subject are in most cases placed together at the end of the relevant paragraph or sentence. Whenever possible, references have been confined to a single minute, memorandum, or correspondence in one file. Reference has otherwise been made either to the number of the jacket of the file containing those entries — particularly in the case of Foreign Office documents — or, where a number of entries in the same file are useful, to the file number if the file is not subdivided into jackets. Unless otherwise specified, all correspondence is in the form of telegrams.

The Wade-Giles system has been used, in preference to the Chinese *pinyin* system, for the transliteration of Chinese terms in order to avoid complicating further the already confusing romanization of Chinese names. Whilst the names of some Chinese leaders have always been romanized according to the Wade-Giles system, the names of others, especially those of Hong Kong or Cantonese origin, have not. Moreover, many of the names have already been used so extensively in government documents and in Hong Kong English-language newspapers that any attempt to standardize their romanized form would increase, rather than decrease, the confusion. In this book, all Chinese names which have already been romanized in the original documents are therefore rendered in the way that they appear in those documents. Names not already romanized are transliterated according to the Wade-Giles system. The same principle has been followed with regard to the titles of Hong Kong Chinese-language newspapers. Although the Chinese practice is to place the surname before other names, this has not been adhered to in the case of those people who chose not to do so — for example, Sir Man-kam Lo (or, simply, M. K. Lo) is not rendered as Lo Man-kam.

2 London Planning

IF British colonial policy before the Second World War was 'certainly not opposed in principle to any idea of eventual self-government it equally certainly did not conceive it to be part of its duty "officiously to strive" to bring self-government into existence'.[1] However, by July 1943, owing to the necessities of a total war against Germany and Japan, British colonial policy had become, in the words of Oliver Stanley (Secretary of State for the Colonies), one of guiding 'colonial peoples along the road to self-government within the framework of the British Empire'. In spite of Stanley's assertion that he had broken no new ground, his statement contrasted sharply with the official position of two years before — that position being the avoidance of fully responsible self-government as the goal for the whole colonial Empire. As such, it was an important development in British colonial policy.[2]

Whilst Britain's colonial policy was being discussed in the Colonial Office, it was also under heavy fire in the Far East. The British defeat in Hong Kong, Malaya, Singapore, and Burma by the Japanese in the winter of 1941 and the spring of 1942 called into question the very existence in the Far East of British colonial administrations. There was also an uneasy suspicion in Britain that defeat had been more than a military failure.[3] This sense of disquiet was increasingly accompanied by a chorus of criticism from the United States. American public opinion had always been anti-imperialist and the British failures had made the British Empire even more objectionable.[4] During the summer of 1942, even the Foreign Office joined the Americans in pressing for an overhaul of Britain's colonial policy in the region. In the light of such pressure, as well as the promises made by the Japanese in colonial Asia, and the fact that an Asian people had achieved supremacy in East Asia, the British government in 1945 reaffirmed 'the very early grant of self-government' as its goal in its Asian colonies.[5]

Hong Kong's fall to the Japanese had prompted the Bishop of Hong Kong, amongst others, to demand radical reform after recovery. Sydney Caine, a senior official at the Colonial Office who had served as Financial Secretary of Hong Kong before the war, considered that if and when the British returned to the colony, the administration would have to be radically changed. The general consensus in the Colonial Office was that the matter should be examined carefully as soon as possible.[6]

In addition to the difficulties shared by all British Asian colonies under Japanese occupation, Hong Kong had a special problem: the Chinese wanted the return of the territory. Whenever possible, China tried to elicit American support, such as at the Pacific Relations Conference held at Mount Tremblant near Montreal, Canada, in December 1942.[7] The Chinese also pressed for a solution to the problem of the future of Hong Kong during the 1942 Anglo-Chinese negotiations for the abolition of extraterritorial rights in China.[8] Although the United States generally endorsed China's claim to Hong Kong, at the critical moment the Americans did not back the Chinese because at that time they themselves were having difficulties with China.[9] The Colonial Office was under such strong pressure in mid-1942, however, that it agreed reluctantly to the cession of Hong Kong, if necessary, as part of a general post-war settlement in the Far East.[10] This being the situation, British officials began to take this factor into serious consideration when examining the constitutional arrangements for the colony.

The uncertainty as to the future of Hong Kong caused considerable anxiety in the Colonial Office. For obvious reasons, it wanted to avoid giving up the colony. In 1943, attempts were made to initiate inter-departmental meetings with the aim of examining the ramifications of the Hong Kong problem but those attempts were frustrated by the Foreign Secretary, Anthony Eden.[11] This failure did not deter the Colonial Office from examining further the future policy for the colony, however, and with the help of the Hong Kong Planning Unit and two confidential advisers from the China Association, the Colonial Office began in early 1945 to tackle seriously both the general problems of post-war Hong Kong and her future status.[12]

The Hong Kong Planning Unit, consisting of a small group of veteran Hong Kong officials, was formed in 1943 as the core of a future civil affairs unit. During its first year, before a permanent Head and Chief Civil Affairs Officer-designate was found, it operated under the wings of the Malayan Planning Unit. When D. M. MacDougall, a veteran Hong Kong Cadet (or Administrative Officer), was appointed Head of the Hong Kong Planning Unit in September 1944, its primary concerns were the preparations for the setting up of an administration and the securing of adequate supplies once the military had recovered the colony. It was also responsible for helping the Colonial Office to formulate future policy.[13] The planners assumed that the war against Japan would not necessarily end with the British return to Hong Kong. Moreover — unlike Singapore and other British eastern colonies — Hong Kong was outside the sphere of Lord Mountbatten's South East Asia Command. It was not known, therefore, whether the colony would be under British, or for that matter, Chinese or American, military command.[14]

On 1 May 1945, unaware of the impending Japanese surrender, the Colonial Office, the Hong Kong Planners, and two civilian advisers from the China Association began to explore officially the possibility of liberalizing the constitution of Hong Kong. The Colonial Office team, led by G. E. J. Gent (Assistant Under-Secretary), suggested that a municipal council be set up and that both the Legislative and Executive Councils be reformed. The central government would maintain control over defence, foreign affairs, finance, the administration of justice, the Police, prisons, labour, fisheries, certain public works, and the duties surrounding the Registrar General's Office. The proposed municipal council would have general authority in matters of a municipal character. It was not entirely clear, however, whether various other government functions would be reserved for the central government or left to the new municipal council. The size of the municipal council would be twenty-three: the European community would elect nine members representing the commercial, professional, and local Portuguese interests; the Chinese community would nominate seven members through their representative guilds; and the remaining seven members, to be nominated by the Governor, would include two representatives of the rural areas. The Legislative Council would be reconstituted to provide a greater representation of the Chinese. Of the fourteen members, five would be official members, two would be ex-officio members (the Chairman of the proposed Harbour Trust and the Chairman of the proposed municipal council), and seven would be unofficial members. Of the seven unofficial members, four would be Chinese, two would be European, and one would be Portuguese. The Governor would remain the Chairman and would have a casting vote. The Executive Council would be replaced by a committee of the Legislative Council with a slightly different title.[15]

While there could be no doubt that the participants in the 1 May meeting, particularly the Colonial Office officials, had given some thought to the problem of reforming the Hong Kong constitution, there is no indication that they had examined fully all of the implications involved. The proposals put forward indicate that there was no question of 'one man, one vote'; the main concern of the planners appears to have been to change the character of the colony in order to placate the anti-colonial American critics, and hence to reduce the anticipated pressure on Britain by the United States to return the colony to China. It is doubtful whether all of the planners were fully aware that reform of the constitution might also put the colony on the road towards further constitutional advance.[16] In any case, this policy of reform was in line with the colonial policy of that time.

A second meeting was held at the end of May 1945 to discuss further the means by which members of the municipal council and the Legislative Council were to be chosen. The planners were faced with a dilemma. On the one hand, it was recognized that the earlier proposal to select Chinese councillors to the proposed municipality (the Chinese not being elected freely by ballot, as were the Europeans) would provoke serious criticism 'in many quarters'; it might also antagonize those who wanted to see the return of Hong Kong to China. On the other hand, the planners objected to 'universal suffrage' because, following the general practice at that time, only British subjects would be given the vote; this would exclude the overwhelming majority of the inhabitants of the colony who were Chinese nationals. The proposed solution to the dilemma was to use the jury list — which included non-British Chinese residents who were literate in the English language — as the electoral list for the proposed municipality. With respect to the election of unofficial members to the Legislative Council, it was agreed after much discussion that the new municipality should act as the electoral body for seven unofficial members, but that the chairman of the proposed municipality should not be given a seat ex-officio. In order to counterbalance the disproportionate influence of the municipal council in the colonial legislature, two additional unofficial members would be nominated by the Governor. The resulting council of sixteen (including the Governor and six official members), would give the unofficial elements a majority of two.[17]

Further consideration of the issue produced two different approaches, one approach being submitted by the Hong Kong Planning Unit and the other by the two advisers from the China Association. Following careful examination of the practical issues involved, the Hong Kong Planning Unit put forward a modified version of the municipal proposals then under discussion. The proposal was prepared by T. M. Hazlerigg, who had retired from the legal Department of the Hong Kong government before the war. Hazlerigg proposed to exclude the rural areas from the scope of the proposed municipality because he thought that the two rural representatives nominated by the Governor would not be able to make their voices heard adequately in a council absorbed in urban and commercial affairs. The number of nominated members should therefore be reduced by two. In view of the size of the local Portuguese community (3,197 in the 1931 census, compared with 6,684 expatriate European British subjects) and the fact that the Portuguese formed the most permanent element of the non-Chinese community in the colony, Hazlerigg suggested that they be given two seats. (Oddly enough, no similar provision was made for the several thousand Eurasians who had also adopted Hong Kong as their home.) Hazlerigg proposed that the Europeans should elect seven members who would represent not only the commercial and professional interests but also the general interests of the whole community.

With respect to the selection of the Chinese councillors, Hazlerigg recognized that nothing short of election by the ballot box would be satisfactory. In the light of experiences in Shanghai, Tientsin, and Kulangsu, he concluded that the nomination of Chinese members by various organizations, instead of election, would not prevent Chinese politics from being brought into the council; however undesirable, the introduction of Chinese politics would have to be expected. Whilst Hazlerigg's concern was justifiable, it remained unclear as to whether Chinese politics would in fact be introduced into the proposed municipality of Hong Kong. The fundamental difference between Hong Kong and the Chinese cities of Shanghai, Tientsin, and Kulangsu was, of course, the fact that Hong Kong was a British colony. In order to prevent impersonation, Hazlerigg also proposed a very elaborate system of identification, which would necessitate different franchises and different electoral divisions for all three communities. His modified proposal provided for a council of twenty-one, to be composed of five nominated members and sixteen elected members, of whom five would be Chinese, two would be Portuguese, and seven would be expatriate Europeans. Hazlerigg's study was based primarily on the pre-war experiences of Hong Kong and other Chinese cities where the British had substantial interests.[18]

Hazlerigg's proposal emphasized that, prior to the war, the system of consultation through the various existing advisory committees and councils had worked well, that there was no serious dissatisfaction with the existing political structure, and that there was no demand for constitutional change. Hazlerigg considered it neither wise nor necessary to link the proposed municipal council to any changes in the power or structure of the Legislative or Executive Councils. Since the municipal council scheme was essentially an experiment, he was anxious to avoid a public commitment of the British government to further constitutional changes in Hong Kong, presumably at least until the results of the experiment were known. He preferred to expand the existing Urban Council gradually into a municipal council by transferring to it various categories of municipal duties. Initially, the council would be responsible for the water supply, drainage, traffic control, roads, building regulations, and the operation of certain hospitals for infectious diseases, the equivalent of the English poor law hospitals. Hazlerigg also pointed out, however, that since the functions of the Hong Kong government were predominantly municipal in character, it was doubtful whether there was room for both a fully fledged municipal council and a central government worthy of the name. Finally, Hazlerigg stressed that no useful action could be taken before full civil government had been restored and the implications of the principles under discussion had been fully analysed.[19]

The proposals in Hazlerigg's memorandum contrast sharply with those hitherto discussed. The original proposals, favoured by Gent, constituted a major reform with significant implications. They introduced indirect elections and an unofficial majority into the Legislative Council, and pointed the way to the introduction of a quasi-ministerial system into the Executive Council. Hazlerigg's proposals, however, confined all the changes at the initial stage to the municipal council; the provisions to change the Legislative and Executive Councils were dropped. In essence, he proposed to expand only the functions of the Urban Council and to widen the basis for local elections — the pre-war Urban Council already had two elected unofficial members. In suggesting this more cautious approach, the Hong Kong Planning Unit had virtually proposed to change the character of the reform.

The unofficial advisers from the China Association, Arthur Morse of the Hongkong and Shanghai Banking Corporation and G. W. Swire of Butterfield and Swire Limited, considered the Hong Kong Planning Unit's proposals to be an attempt to preserve intact the whole colonial structure by creating a 'camouflage' municipal council — a council with none of the functions of a municipality, but with only extended duties of the old advisory committees and of the Urban Council. They considered this to be the wrong approach and insisted that a full municipal council be set up to deal with all municipal affairs, leaving the Governor and the Executive and Legislative Councils free to handle other non-municipal issues. Morse and Swire also stressed that unless the council were to be given a sufficiently wide range of responsibilities, influential, able, and fully competent Chinese would not be attracted to serve on it. They disagreed with Hazlerigg's suggestions with respect to the selection of Chinese councillors, preferring that they be selected by various representative Chinese organizations. Even more so than Hazlerigg, Morse and Swire referred to the experiences in pre-war Shanghai in order to support their proposals and arguments (which were prepared by a former Secretary of the Shanghai Municipal Council, J. R. Jones). However, they agreed with Hazlerigg on the need to proceed cautiously.[20]

The Colonial Office agreed with the view of the China Association representatives that the Hong Kong Planning Unit's proposals represented less of an advance than was desirable. The Colonial Office wanted a bold approach, including the transfer to the proposed municipality of a sufficiently wide range of functions to attract responsible Chinese to serve on the council. Although the Colonial Office disagreed on the subject of franchise with the more restrictive proposals of the China Association, it was also dissatisfied with the Hong Kong Planning Unit's suggestion.[21] These deliberations were disrupted abruptly, however, when Japan suddenly surrendered on 14 August 1945 and members of the Hong Kong Planning Unit were given military ranks and dispatched to the colony to take control of civil affairs.

British planning for post-war constitutional development in Hong Kong was not undertaken only in London. F. C. Gimson, Colonial Secretary of Hong Kong prior to the Japanese occupation and subsequently the senior British official in the colony, started parallel planning on his own initiative inside Stanley Internment Camp, Hong Kong, in around July 1943.[22] These deliberations were not completed before the Japanese surrender and the records were destroyed. Gimson and his Executive Committee examined the problems of post-war constitutional reform, including the representation of the Chinese and the development of local self-government. Although Gimson recognized the danger that the Chinese government might try to interfere in local politics, he favoured the introduction of 'popularly elected' Chinese representatives into the Legislative Council. It was also unanimously concluded that two municipalities should be set up, for two reasons. First, the requirements and size of the two cities (Victoria and Kowloon) were very different. Secondly, Gimson and his Committee considered that two smaller municipalities, even if elected, would have less claim as representing the people in disputes with the central government than one large municipality. No agreement was reached, however, on the question of franchise.[23]

Gimson considered himself to be the leading light in the planning of Hong Kong's constitutional development; he considered the traders to be merely 'birds of passage', who regarded Hong Kong in the same way as they regarded the treaty ports of China. They had no concern for the colony as such and no appreciation of the 'imperial concept of citizenship'; they were unwilling to discuss Hong Kong's commercial development, let alone its political development. The civil servants were also, in general, 'woefully backward' in their attitude towards colonial political development. Indeed, those most strongly opposed to the introduction of self-governing institutions in accordance with the British government's colonial policy were Gimson's fellow officials. R. A. C. North, Secretary for Chinese Affairs, for example, openly contradicted and opposed Gimson on the subject. Regretfully, North's arguments cannot be traced.[24]

Just prior to the British reoccupation of Hong Kong and the setting up of a military administration in September 1945, a serious dispute over the colony broke out between Britain and China. According to the wartime arrangements of the Allied

Powers, Hong Kong was within the China theatre, which was under the overall command of Generalissimo Chiang Kai-shek, the Supreme Commander of the Allied Forces in the China theatre. It was therefore not surprising that Chiang expected the liberation of Hong Kong to be carried out by forces under his command. However, Britain was determined to take over from the Japanese in Hong Kong for three reasons. In the first place, the British services felt it to be a matter of honour that, having lost the colony to Japanese arms, they should recover it directly from Japan. Secondly, it was hoped that a British military reoccupation of the colony would to some extent restore British prestige in the Far East, which had been shattered by her defeats there in 1941–2. More importantly, Britain wished to avoid a Chinese military occupation of Hong Kong, which might considerably strengthen the Chinese position in any negotiations over the future status of the territory. The dispute came to a head when Britain simply notified Chiang Kai-shek of her intention to send a naval detachment, led by Rear Admiral C. H. J. Harcourt, to recover Hong Kong from the Japanese. As Supreme Commander, Chiang felt that he should have been consulted beforehand and that the order to Harcourt should have been issued by him. He considered the British action to be an insult, and refused to allow any British forces not under his command to take over Hong Kong. As the dispute continued, both Chiang and Prime Minister Winston Churchill appealed to President Truman of the United States for support. Eventually, after Truman sided with Churchill, Chiang backed down and consented to the recovery of Hong Kong by Harcourt's British naval task force before the end of August. During the course of the dispute, British officials thought that Chiang harboured sinister motives with regard to the colony. In retrospect, however, it would appear that Chiang's main concern was 'face'. In any case, the dispute undoubtedly reminded the British officials responsible for Hong Kong of the need to strengthen Britain's position in the colony vis-à-vis Chinese irredentism over the territory.[25]

Following the recovery of Hong Kong, the Colonial Office resumed its examination of a constitutional policy for post-war Hong Kong, but this time without the advice of the Hong Kong Planning Unit or the China Association. By this time, the Conservative Party had been defeated in the British general election of July 1945, and Oliver Stanley had been replaced as Secretary of State by G. H. Hall of the Labour Party. Since the Labour Party at that time was relatively more committed than the Conservative Party to leading British dependencies to greater self-government within the Empire, a more favourable general atmosphere prevailed for introducing a policy of reform in the colonies.

In spite of the general atmosphere, however, Hall's term as Secretary of State had no immediate impact on British policies for Hong Kong. Indeed, at that time, the main advocate of the proposed reforms for Hong Kong was Gent, the Assistant Under-Secretary of State supervising Far Eastern affairs in the Colonial Office. Gent was very keen to reach an early decision on the matter. He recognized that it would be neither possible nor desirable at that stage for London to settle all of the details, but he stressed the need for a general statement of the British government's intention to give the inhabitants of the colony a fuller and more responsible share in the administration of their own affairs, even before the restoration of civil government. Gent emphasized that the people of Hong Kong expected more than the restoration of the *status quo ante bellum*, and that official and unofficial opinion favoured an extension of democratic government in the new era. The climate of opinion indicates that Gent was not interested in reform merely for the sake of silencing American and Chinesé critics. He apparently felt that domestic considerations alone in Hong Kong were sufficiently important to put the colony on the road to constitutional advance. Unlike the Hong Kong Planning Unit, Gent wished to reform the Legislative Council as well as to establish a municipal council. In essence, he revived the 1 May proposals which had advocated the introduction of a municipal council of twenty-three members (consisting of nine elected European members, seven indirectly elected Chinese members, and seven nominated members) and a Legislative Council of fourteen members (consisting of seven unofficial members and seven official members). The municipal council was also to have authority over the rural areas.[26]

Gent's approach should not be seen simply as an attempt to introduce elected local government to Hong Kong. Since the jurisdiction of the proposed municipality would be coextensive with the area of the colony, what was being proposed was a form of diarchy — a system in which certain functions of government are administered under a separate organization with its own legislative body formed on a different constitutional basis from the colonial legislature which controls the remainder of the administration.[27] The merit of such an approach is that the British government could use the proposed municipality as a training ground for democratic self-government for the local people. If the experiment proved to be a success, and if the British government so decided, the municipality could in due course replace the colonial government. If the experiment were to be unsuccessful, however, the colonial government presumably could resume various functions from the municipality without causing a major constitutional crisis in the colony. This two-tier approach also had the advantage of meeting the contradictory requirements of reform of the constitution of Hong Kong: the need to restrict participation in the colonial legislature to British subjects; and the need to satisfy Chinese and American opinions by the introduction of election by ballot at the municipal level to the Chinese inhabitants of the colony. Although Gent did not point out these merits of the two-tier approach, it is unlikely that he was unaware of them.

Although Gent did not say explicitly that reform was also desirable in order to strengthen the British position in the event of a demand by China for the retrocession of Hong Kong, it must have been an important consideration: various papers dealing with both subjects were filed together and he was at the time supervising the examination of both problems.[28] Furthermore, his wartime duties and attitude suggest that he had always linked the two issues. As the Assistant Under-Secretary responsible for both Hong Kong affairs and public relations during the war, Gent defended the British Empire against American criticism. He also defended Hong Kong as an integral part of the Empire. When the Foreign Office had recommended in the summer of 1942 that Hong Kong and other 'non-essentials' be given up in order to protect the 'really important things', Gent had strongly deplored the treatment of British territories as if they were pawns in the game of international politics.[29] He subsequently attempted to initiate interdepartmental studies into the ramifications of the Hong Kong problem.

In general, the Foreign Office linked the reform proposals to the future of the colony. It regarded efforts of the Colonial Office as attempts to 'stave off the evil day'. While Foreign Office officials conceded that reform was a step in the right direction in

accordance with the spirit of the times, they remained sceptical as
to its effectiveness. Nevertheless, they agreed that the inhabitants
of Hong Kong should not be denied a measure of self-
government simply because it would not go far enough to satisfy
the Chinese, who wanted the return of the colony. The Foreign
Office was very concerned about the problem of Hong Kong's
future status. A Chinese demand to begin discussions on the
future of Hong Kong's leased territory was expected: Britain had
agreed in early 1943 that the Chinese might raise the issue after
victory had been achieved, and the Chinese government had
indicated that it wanted the return of Hong Kong.[30] Under such
circumstances, the Foreign Office was anxious to project a new
image. Any indication that the *status quo ante* was to be restored
was to be avoided. For this reason, it objected to Gent's use of
the term 'colony' in his draft statement on reform for Hong
Kong. It also favoured changing the Governor's title, and fought
very strongly (though unsuccessfully) against the reappointment
to the Hong Kong governorship of its pre-war incumbent, Sir
Mark Young.[31]

In January 1946, Young (who had been imprisoned in China by
the Japanese during the war) began to take part in the reform
deliberations. His first reaction was doubt as to the wisdom of
concentrating on the expansion of the functions of the proposed
municipality. After further study, he suggested some minor
amendments to Gent's draft statement of the British government's intentions. Young strongly urged that the matter not be
settled in London before the Governor could consult fully with
the responsible leaders of the community.[32] In so doing, he
avoided committing the Hong Kong government to any specific
course of reform before a full assessment of the local situation
could be undertaken.

Prior to the restoration of civil government in Hong Kong in
May 1946, Whitehall had reached a conclusion on the principle
governing reform in the colony: the general policy of providing
colonies with a greater measure of local self-government should
also be implemented in Hong Kong. However, the means by
which this was to be achieved had not been decided. (The Colonial Office preferred the development of local government, but it
would be willing to consider a continuation of the centralized
arrangement, coupled with a liberal franchise and an unofficial
majority in the legislature.) The decision would be made after the

Governor had assessed local public opinion. In the meantime, it was also decided that the local inhabitants should be given the opportunity to enter the public service.[33]

The Hong Kong Scene

As soon as the Japanese surrender was announced, the British government sent instructions to Gimson, the senior British official in the colony, through a Chinese agent of the British Army Aid Group which operated in south China. These instructions stressed the British government's decision to restore sovereignty over, and its administration of, Hong Kong, and instructed Gimson to assume power as the Officer Administering the Government, pending the arrival of British forces which would set up a military administration.[34] Before the instructions had been received, however, and with an eye to the future status of the colony, Gimson had already demanded that the Japanese help him to re-establish British civilian rule. He ordered the Japanese to keep law and order, while he restored a skeleton administration on the 1941 basis, with officials drawn from the Stanley Internment Camp and from various prisoner-of-war camps.[35] On 1 September, the day following the British Fleet's reoccupation of Hong Kong, Rear Admiral Harcourt established a British Military Administration and appointed Gimson Lieutenant-Governor by proclamation. A curious situation resulted, in which a military administration with full power co-existed with a civilian government advised by an Executive and a Legislative Council. The situation was later rectified when Gimson's appointment as Lieutenant-Governor was disallowed and his administration was merged with MacDougall's Civil Affairs Branch of the Military Administration.

In spite of the initially anomalous and confusing constitutional structure, the British authorities very quickly began to tackle the most important problems of the day. The immediate military problems were relatively simple: to take over control and general Police duties from the Japanese, and to disarm and intern them as soon as the necessary military forces were available. The civil affairs problems were more difficult to resolve. The establishment by MacDougall of an administration was both helped and hindered by Gimson's earlier efforts. While there was no need to establish an administration from scratch, Gimson's skeleton administration matched neither the blueprint worked out in London by MacDougall and his team nor the exact requirements of the day. The problems on the administrative side were so complex that MacDougall thought even an army of professional 'fixers' would have been unable to straighten them out by 1947. Instead, MacDougall and his team turned their attention to the immediate problems: currency, labour, public health, food, and fuel.

Under MacDougall's able leadership, these problems were handled with vigour and often regardless of rules, or regulations, or pre-war conventions. In order to restore the economy on a sound basis, the Military Administration promptly substituted the Hong Kong dollar for the Japanese military yen, even though the quantity of dollars available was very small. A moratorium was also imposed on all debts incurred during the occupation. Banks and utility companies were given assistance to enable them to resume operation as soon as possible. A government reconstruction scheme averted the immediate threat of labour and general unrest and the jobless were employed to clear the streets and to repair war damages. In addition, the price of food and other essential goods was controlled, relief was provided to those who needed it, and ships were sent to Chinese and South-east Asian ports for food and fuel.

A significant and fruitful attempt at self-help was initiated by G. A. C. Herklots, a biologist at the University of Hong Kong. Herklots devised and, with the assistance of Jack Cater, an officer of the Royal Air Force, implemented a fishing co-operative scheme. The scheme, which had the support of the administration, immediately made it possible for Hong Kong's fishing fleet to put to sea, thus securing for the colony an important source of food. Herklots and Cater also later introduced a similar agricultural co-operative project. The remarkable long-term achievement of these schemes was their ending of the traditional exploitation of fishermen and farmers by the *laans* — wholesaler and loan-shark combined — who had previously had firm control over those industries. In spite of the severe and widespread shortages, Hong Kong managed to keep one step ahead of disaster. In November 1945, for example, the stock of food reached the dangerously low level of only ten days' supply. Before the end of the year, however, the banks were fully operational, public utilities were running, and the whole colony was reopened to trade. Hong Kong was once again bustling with life and some of the worst fears of its inhabitants were over.[36]

Immediately following the Japanese surrender, the Chinese inhabitants of Hong Kong viewed the return of the British with both relief and indifference. Unlike the British officials, the local inhabitants were unaware of the Chinese government's attitude towards the future status of the colony. The British had always been in Hong Kong; their return was logical and to be expected. After the brutality and hardships of the Japanese occupation, the local Chinese inhabitants remembered the pre-war British administration as being benevolent and efficient. Notwithstanding this attitude to the British, in the city itself four times as many Chinese national flags as Union Jacks were displayed. In the euphoria of victory, the local people identified themselves with China. For them, as for the rest of the Chinese nation, the victory over Japan was a great moment: it marked not only the end of the war and of misery, but the beginning of a new era — one in which China would be recognized as one of the five Great Powers.

After the initial euphoria had subsided and some of the most pressing problems had been dealt with, the people of Hong Kong developed what Harcourt called the '1946 outlook'. This outlook consisted not only of a sense of national pride in China, but also of a feeling that the *status quo ante* was anathema. Chinese national pride manifested itself not only in the displaying of flags, but also in the columns of local newspapers and in the public reactions to local incidents. The leading local newspapers occasionally reminded the local people to behave in ways befitting the citizens of a Great Power.[37] There was a strong and swift reaction to the death of a Chinese girl at the hands of a British rating in early October 1945 and to similar incidents involving Chinese and Europeans. Two local riots broke out when a British (in fact, Indian) police constable on street duty inadventently caused the death of a peanut hawker. Significantly, the temper of the crowds was anti-European. In sharp contrast to the pre-war days, a slap in the face of the Chinese would no longer be tolerated.[38] This change became an important factor in the socio-political scene of post-war Hong Kong.

Once the immediate memory of the harshness of life under the Japanese had receded, the attention of the local people turned to conditions as they had been under the pre-war administration.

Prior to the war, there had been too much privilege, snobbery, discrimination, race prejudice, corruption, and absentee exploitation. The greatest failure of the pre-war government was seen to be its disregard for the local people's interests. Considered to be essential in 'new Hong Kong' were better and fairer treatment of the Chinese; carefully chosen, incorruptible British officials, with some knowledge of the Chinese; free education for the children of the poor; and, most importantly, an almost complete absence of the 'old gang'. The return of the pre-war officials was perceived as a potential threat to the new order. Some of those officials who had not divested themselves of the 'treaty port mind' were in fact keen to restore the old order. The articulate public made it clear, however, that any attempt to resuscitate the worn out government machinery which had failed them so badly in 1941 would not be acquiesced in meekly. They looked forward to a radical reform which would provide the people of the colony with a greater say in public affairs. They wanted the framework of the new constitution to be clearly defined so that the new civil government would be built on such a basis. They were not specific, however, as to what they wanted in the new constitution.[39]

This new '1946 outlook' was keenly observed and reported to Whitehall by far-sighted British officials and non-officials. Both Harcourt, Head of the Military Administration, and MacDougall, Chief Civil Affairs Officer, independently urged the Colonial Office to arm the new civil Governor, at the time of the restoration of civil government, with a programme of reform, or at least an announcement of an outline for reform. John Keswick, *taipan* of Jardine, Matheson and Company and a wartime political adviser at the Chungking Embassy, went a step further. He strongly urged the British government to transform the Crown colony of Hong Kong into what he called the 'Free Port and Municipality of Hong Kong'. Keswick thought that the Governor's title should be changed and that he should be assisted by an elected council. 'Crown colony', 'Governor', and 'selected councils' were unacceptable terms in the new era. Keswick was concerned about more than the changes that had already taken place in Hong Kong; he was also very aware of the uncertainty over the future status of the colony, stating that reform was necessary both to fulfil the British policy of leading colonies to self-government and, more significantly, to take some of the wind out of the sails of the Chinese and American critics. With an eye to relations with China, he also proposed that the Secretary for Chinese Affairs be replaced by a Secretary for Chinese and External Affairs.[40]

The beginning of the British Military Administration also saw some fundamental changes in the nature of Chinese political activities within the colony. The main changes, which were caused by the relative decline of British power and the rise in the prestige of China which resulted from the war and the Japanese occupation, were the acceptance by Britain of official Chinese government representation, and the open as well as legal existence of the Kuomintang in the colony.

Shortly after the Japanese surrender, China and Britain negotiated the appointment for the first time of a Chinese Consul-General in Hong Kong. When T. W. Kwok was appointed in November 1945 to represent China officially, his title was not 'Consul-General', but 'Special Commissioner for Hong Kong'.[41] The choice of title was important. Although Chiang Kai-shek's government never officially repudiated British sovereignty over Hong Kong, the appointment of a Chinese Consul-General would have formally sealed Chinese acceptance of British sovereignty — something the *Wai-chiao-pu* (the Chinese Foreign Ministry) wished to avoid. The appointment of Kwok to this special position, in addition to his appointment as Special Commissioner for Kwangtung and Kwangsi, would, on the contrary, convey a different meaning. The combining of these two appointments with similar titles in the person of Kwok meant that the Chinese government did not distinguish Hong Kong — Chinese territory temporarily under British administration — from Kwangtung and Kwangsi provinces — undisputed Chinese territories. The choice of the preposition 'for' rather than 'in' in Kwok's title was also significant. Whereas a Special Commissioner *in* Hong Kong would be interpreted to mean the Chinese government representative in Hong Kong, a Special Commissioner *for* Hong Kong could be interpreted to mean the Chinese government official responsible for the territory. This ambiguity could therefore be of use to China in future negotiations with Britain. In any case, Kwok was looked upon by many local Chinese as the head of their community, and his office became the focus of local attention whenever an incident occurred involving the Chinese and non-Chinese. Kwok's presence also challenged in some ways the hitherto undisputed leadership in the local Chinese community of the appointed Chinese members of both the Executive and Legislative Councils.

The Kuomintang had been outlawed in Hong Kong before the war. After the Japanese occupation, however, it had to be accepted, having built up a large underground organization in the colony during the war. It was heavily armed and was deemed difficult to expel without considerable bloodshed and disruption of the Administration's work.[42] The crucial factor, however, was the fact that the Kuomintang was the ruling party in China and the necessities of foreign policy in the light of the post-war situation ruled out rough handling of the party in the colony.

With the exception of a few incidents in which local Kuomintang supporters were involved, the Kuomintang co-operated with the British authorities in Hong Kong. When Sum Chit-son, the Kuomintang representative in the colony immediately after the war, organized without notifying the Hong Kong Administration a series of demonstrations to honour General Pan Hwa-kuo, the head of the Chinese Military Mission in the colony, the British Minister in Chungking (G. A. Wallinger) took the matter up with General Wu Te-chen, the Kuomintang Secretary General. Wu heeded the British representation by promptly replacing Sum.[43] The Kuomintang also refrained from inciting labour unrest when they could easily have done so, and generally co-operated with the British military authorities.[44] Given the fact that more than 100,000 Kuomintang troops passed through the colony on their way to reoccupy north and north-east China, it was remarkable not that there were some incidents involving Kuomintang soldiers but that the incidents were so few in number and so quickly defused.[45] If the Kuomintang had wished to make things difficult for the British authorities, they could have done so. Despite occasional excesses by some local zealots, the Kuomintang appears on the whole to have been reasonably well disposed towards the Hong Kong Administration. Moreover, MacDougall, the Chief Civil Affairs Officer, was not alarmed by the effect of Kuomintang activities on the reform proposals. He stressed that, compared with the immediate pre-war days, the Chinese Military Mission, which was considered by the British Minister in Chungking to be large, was a mere trickle.[46]

The other important aspect of Chinese politics in post-war Hong Kong was the activities of the Chinese Communists. During the war, the Communist East River Column had set up an efficient underground network in and around Hong Kong. It had also maintained very good working relations with the British Army Aid Group.[47] Immediately after the war, Communist guerrillas maintained law and order on various islands and in parts of the New Territories until the British commandos were able to assume command. The bulk of the Communist forces subsequently left the colony, though they left behind them small political nuclei.[48] These groups of political partisans were gradually reinforced by senior cadres from Yenan whose main task in the colony was to rally Chinese support, through propaganda and other means, for the Communist struggle for power on the Chinese mainland. As part of this united front work, the Communists' most important party propaganda machine, the *Hwa shiang pao* (*Chinese Merchants' Daily*), was set up in Hong Kong to print a newspaper primarily for circulation on the mainland.[49] Hong Kong was, for the Chinese Communists, merely a safe sanctuary from which to operate in south China.

Throughout the British Military Administration, the Chinese Communists adopted a very low profile: they provoked no incidents, and were not regarded as trouble-makers by the British authorities. Nevertheless, they were perceived as *potential* mischief-makers and were watched carefully. The results of this observation were reassuring to the British: local Communist propaganda was concerned primarily with events in China; when it did refer to the British, it did so in very friendly terms.[50] There appeared to be no need at that time for the British authorities to feel apprehensive about their activities.

By the end of April 1946, when the Military Administration was about to be replaced by a civil government, the immediate post-war problems had largely been solved. Supply was still a problem, but this was the situation world-wide. Allegations in the early days of collaboration with the Japanese had been potentially disruptive, but those had died a natural death.[51] The population had increased from less than 600,000 when the British returned, to well over one million by early 1946. However, this increase represented mainly the return of residents who had been forced to leave during the Japanese occupation rather than an influx of newcomers or refugees.[52] The herculean task of general rehabilitation and reconstruction of the economy and society of Hong Kong had barely begun, but this was a long-term problem for the post-war civil government. On the whole, the Military

Administration, as a stopgap government, had done very well in facilitating the transition to civil government. Harcourt considered the main problems left by his administration for the civil government to be the introduction of a new constitution; the placing of Chinese into positions of responsibility in the government; the elimination of the colour bar; and the return of some old-timers who failed to realize the need for a new '1946 outlook'.[53]

3 The Young Plan

On 1 May 1946, Sir Mark Young resumed his interrupted governorship and restored civil government in Hong Kong. On this occasion he declared:

His Majesty's Government has under consideration the means by which in Hong Kong, as elsewhere in the Colonial Empire, the inhabitants of the Territory can be given a fuller and more responsible share in the management of their own affairs. One possible method of achieving this end would be by handing over certain functions of internal administration, hitherto exercised by the Government, to a Municipal Council constituted on a fully representative basis. The establishment of such a Council, and the transference to it of important functions of government might, it is believed, be an appropriate and acceptable means of affording to all communities in Hong Kong an opportunity of more active participation, through their responsible representatives, in the administration of the Territory. But before a decision is taken on the methods of giving effect to the intentions of His Majesty's Government, it is considered essential that the important issues involved should be thoroughly examined in Hong Kong itself, the fullest account being taken of the views and wishes of the inhabitants. The Governor has accordingly been instructed to examine the whole question, in consultation with the representatives of all sections of the community, and to submit a report at an early date, bearing in mind the policy of His Majesty's Government that the constitution should be revised on a more liberal basis as soon as possible. The aim will be to settle and to announce not later than the end of the year the principles on which that revision should be based.[1]

This statement, which was also made simultaneously by Secretary of State George Hall[2] in the House of Commons, committed the British to changing the constitution of Hong Kong, and defined the direction, scope, and procedure of reform. It is significant that the object of reform was to give the local inhabitants — not just the British subjects — more substantial participation in the management of their own affairs through their responsible representatives. The statement also committed the government to consulting fully all sections of the local community and to settling the governing principle for a liberal reform before the end of 1946.

Young commenced his task by appointing T. M. Hazlerigg as Special Adviser and a member of the Executive Council. Hazlerigg was to assist Young in the collection and collation of the views of the community, as well as in the formulation of his recommendations. Before the end of May, Young invited various representative bodies to give their comments and views on some of the questions involved in the proposed revision of the constitution. Members of the public were also asked to express their opinions, and a questionnaire on local self-government was published in two of the local English-language newspapers. On 28 August, at the completion of the first stage of consultation, Young made a radio broadcast in both the English and Chinese languages to the people of Hong Kong.[3]

In his broadcast, Young outlined his initial findings and preliminary conclusions as a basis for further discussion. He considered a municipal council for the urban areas of the colony the best means by which to achieve the objects set out in the 1 May statement. He proposed a council of forty-eight, constituted of sixteen elected Chinese members, sixteen elected non-Chinese members, and sixteen nominated members (half of whom were to be nominated by Chinese bodies and half by non-Chinese bodies). The council would elect its own paid chairman, enjoy a high degree of financial autonomy within the framework of its constitution, and provide ample opportunities for locally recruited officers. In the first instance, it would manage parks and playgrounds, taking over the power of the existing Urban Council, but it would gradually assume responsibility for public sanitation, education, social welfare, building, town planning, the supervision of public utilities, general licensing, and the Fire Brigade. Apart from the general literacy and property qualifications, Chinese voters would be required to have lived in the colony for six out of the last ten years, whereas the requirement for non-Chinese would be residence for only one out of the last five years. There would also be separate elections for the Chinese and the non-Chinese elected councillors. Young stressed that these conclusions were merely preliminary, and he invited comments and discussion.[4]

After the broadcast, Young discussed with representative individuals and associations both the general question of constitutional reform and the specific proposal of setting up a municipal council. In the light of these discussions and comments which appeared in the newspapers, Young modified and developed further his earlier proposals before making his report to the Secretary of State in October.

Young's principal recommendation was the establishment of a municipal council of thirty members for Hong Kong island, Kowloon, and New Kowloon, excluding the larger part of the mainly rural New Territories. The council would consist of equal numbers of Chinese and non-Chinese councillors, chosen on the following basis: one-third would be elected directly by the Chinese; one-third would be elected directly by the non-Chinese; and the rest would be nominated by selected responsible bodies representing both communities. The council would have slightly wider functions than those outlined in the broadcast of 28 August. Initially, it would be responsible for the Fire Brigade, parks, gardens and recreational grounds, the licensing of vehicles, the licensing and control of places of amusement, and all of the functions of the Urban Council. In due course, it would assume responsibility for education, social welfare, public works (water supply, roads, drainage, and buildings), the supervision of public utilities, and the granting of franchises for them. Prior to the transfer of these functions, the complicated issues involved would be studied by a special commission.[5]

The qualifications for voters would be a minimum age of twenty-five; literacy in either the English or the Chinese language; either eligibility to serve as a juror or the ownership or rental of a piece of property on which rates of at least $200 per annum were payable; and a period of residence in the colony of six out of the last ten years (for those Chinese who were not British subjects), or one year since the attainment of the age of twenty-three (for British subjects). Membership of the council would be open to voters who had the following additional qualifications: a minimum age of thirty; the ability to speak, read, and write English; connections with the area to be represented; and (for the non-British Chinese residents) a period of residence in the colony of ten out of the last fifteen years. There would be separate electoral registers for the Chinese and the non-Chinese voters. There would be six wards on Hong Kong island and four on the mainland for the Chinese electorate. With respect to the non-Chinese electorate, the entire municipality would form one constituency with ten seats, with a special provision to ensure that there would be at least one representative each from the

Portuguese and Indian communities. The representative bodies which would be responsible for nominating the remaining ten nominated seats would be the Hong Kong General Chamber of Commerce (two non-Chinese), the Chinese Chamber of Commerce (one Chinese), the Unofficial Justices of the Peace (one Chinese and one non-Chinese), the University of Hong Kong (one Chinese), the Hong Kong Residents' Association (one non-Chinese), the Kowloon Residents' Association (one non-Chinese), and the recognized trade unions (two Chinese).[6]

Young also stressed that the proposed municipal council should enjoy complete control over its own affairs, including municipal finance and the appointment of municipal officers. It should not be subject to any form of external control (that is, there should be no special power reserved to the Governor or the colonial legislature).[7] What Young proposed to set up in practice was not an ordinary municipal council after the British model at that time. As a local authority, a municipal council in Britain was subject to a certain amount of ministerial supervision and financial control by Westminster (or, in a colonial context, by the colonial government and legislature). Young wished to exempt the Hong Kong municipality from such external control. In his view, such autonomy was necessary because the Hong Kong municipality was to be more than just a local authority; it was to be the training ground for turning the Chinese and other residents of the colony into citizens of British Hong Kong. In this respect, Young's recommendation echoed Gent's earlier suggestion of setting up, in practice, a diarchy. Whether this consideration could justify a departure from the usual practice was, of course, a different question.[8]

Young's second recommendation was the modification of the Legislative Council in order to secure the increased and more direct representation of the Unofficial element. In view of the transfer of certain duties to the municipal council, the size of the Legislative Council would be reduced from seventeen to fifteen seats. An Unofficial majority of one would be secured by reducing the number of official members by two. The new Council would therefore have five ex-officio, two official, and eight unofficial members. The pre-war practice whereby both the Hong Kong General Chamber of Commerce and the Unofficial Justices of the Peace nominated two unofficial members to the Council would be formally accepted and extended. Two of the other unofficial members would be nominated by the new municipal council.[9]

In making the above recommendations, Young stressed that although the proposals could be regarded as carrying the assent of most of those who had displayed an interest in the subject, they could not be said to represent 'even the strongly expressed desire of any large section of [the community]'. His inquiry was met with little enthusiasm. Indicative of this apathy were the 'insignificant' number of members of the public who submitted their views, the relatively small coverage of the subject in the local newspapers, and comments made by more than one public body about the indifference of its members.[10]

Young also attributed the lack of enthusiasm for internal political change to apprehension, which he felt was composed of two elements. First, it was believed that any transfer of power to representatives of the people, whether nominated or elected, would strengthen the hold of the Kuomintang over the institutions and people of the colony. There was even some fear that a municipal council might be used by the Kuomintang to declare itself in favour of the retrocession of Hong Kong to China. Young thought that there were good grounds for such a fear: 'Chinese political parties would seek to use the Municipal Council for their own ends.' Although Young did not specify the people who held these views, they were probably the rich and educated who were close to the colonial government, rather than the local Chinese residents, the vast majority of whom were struggling to make a living, with little time to spare for such matters. The other cause for concern was the uncertainty over the future status of the colony. Many local Chinese regarded it as possible (some even thought it certain) that the British government would be unable to withstand the Chinese demand and would retrocede the colony to China in a few years' time. It was therefore necessary to be cautious in their political conduct lest their 'open collaboration' with the British authorities — in a reform which indicated the British intention to stay in Hong Kong — became a cause for later punitive action. According to Young, it would be more expedient, even for the common people, to 'keep in with both sides' and do nothing which could later be construed as being anti-patriotic to China.[11]

Young also reported the fear in the community that municipal government might lead to corruption and wholesale jobbery. He admitted that there was a risk of corruption in the council, but he also pointed out the widespread corruption even under the existing system of government. In his view, corruption was not a sufficient reason for withholding the opportunity for political advance; on the contrary, popular participation in public affairs would probably create some sense of civic responsibility, which would improve public morality.[12]

In spite of (or, indeed, because of) these observations, Young considered his two recommendations to be appropriate and complementary in the circumstances, and that political apathy 'manifestly needs to be overcome by political education and by an insistence on the transfer of responsibility'. The object of reform was to help Hong Kong 'develop an active sense of citizenship' and to enable its inhabitants openly to express and give practical effect to their general desire 'to remain under British rule and to resist absorption by China'. Young thought that the best countermeasure to Kuomintang infiltration was the framing of the constitution of the municipality in such a way as to preclude it from concerning itself with political matters, particularly the future status of the colony. In part to pre-empt a propaganda attack by the Kuomintang, Young insisted that the municipal council be opened to non-British Chinese residents. The changes to the Legislative Council were seen as essential because they would 'give satisfaction in most quarters'.[13]

Young's approach rejected the logical alternative of concentrating on the reform of the colonial legislature. Such an option would involve dealing with the question of whether the vote should be restricted to British subjects and Young found this question intractable. If the vote were to be restricted to British subjects, it would mean that participation would be confined to a maximum of twenty-one per cent of the population (that is, those who could claim British nationality because of local birth), which would hardly be in line with the 1 May statement. The opening of the Legislative Council to non-British Chinese residents, however, would mean that the feelings of the loyal British subjects — who were strongly in favour of the taking of the oath of allegiance — would be overridden. Moreover, such a change would require a departure from the usual practices in British colonial legislatures. Potentially far more damaging was the danger that it would also open the Legislative Council to Kuomintang infiltration. Unlike a municipal council, the Legislative

Council could not be prevented from becoming involved in political matters. In any case, Young preferred to establish a generally elected municipal council with full power rather than a partially elected Legislative Council with only advisory committees studying selected departments.[14]

Young did not refer in his report to the Secretary of State to the pre-war experience of the Singapore Municipal Council or to the parallel constitutional developments then under consideration in Singapore. He explained later that this omission was because he did not think the Singapore model relevant. First and foremost, Singapore's heterogeneous population was not subject to the same kinds of political influences which the major political parties of China exercised in the predominantly Chinese community of Hong Kong. In addition, and in contrast to Hong Kong, the Singapore population was made up by non-transient residents who had had previous municipal experience.[15]

In spite of the differences in the size of the proposed council and other details, Young's scheme was in essence similar to that put forward by Gent in September 1945. Both proposals envisaged the introduction of what was in effect a two-tier system of government. As explained in Chapter 2, a particular advantage of this approach was that if the democratic experiment with the municipality were to fail, the colonial government would be able to take over without causing a major constitutional crisis. However, in order for this to be the case, it was essential that the unofficial members of the legislature, or at least the vast majority of them, should not be drawn from the municipality. If that were to be the practice, any attempt to abolish the municipality might require that the means of selecting the unofficial members (or a majority of them) to the legislature be changed, leading possibly to a constitutional crisis. Herein lay an advantage of Young's scheme over Gent's proposals. Whilst Gent had not ruled out the possibility of allowing the municipality to elect most of the unofficial members to the legislature, Young limited the number to two. Since the elected municipality was to be an experiment, it would have been better not to have linked it to the Legislative Council at all, but not to have done so might have led the local people to view the municipality as a sham, which would have been the surest way to ruin the experiment. The conflicting requirements outlined above indicate that Young's scheme was not only unavoidable but also probably the most appropriate solution in the circumstances.

The first reaction of the Colonial Office to Young's recommendations and observations was to compare them with the professed aims of the British government, with the conclusions reached earlier in the Office, and with similar developments in Singapore. The present proposals were seen as less drastic than might have been expected of a bold move to provide a measure of self-government. In view of the practical considerations which Young had put forward, however, the recommendations were regarded on the whole as well balanced.[16] Young's case for not having referred to the Singapore experience was also later accepted as valid.

Further examination of Young's proposals produced three main lines of thought amongst the officials in the Colonial Office. The head of the Hong Kong Department, N. L. Mayle, who had little experience of Far Eastern or Hong Kong affairs, strongly opposed the municipal scheme. He referred to Young's warnings about the apathy of the Chinese inhabitants and the risks involved. He felt that the holding of elections would cause 'serious disorder and disturbances'. He also doubted whether the Chinese would elect the ten Chinese councillors as required in the proposals and he suggested that the municipal council be filled with nominated members at an early stage.[17] Mayle overlooked the fact that even in the pre-war Urban Council there had been two directly elected unofficial members, and that the pre-war elections had not caused 'serious disorder and disturbances'. Essentially, Mayle wanted no reform in Hong Kong unless it was politically necessary.[18]

Another approach to the Hong Kong reforms was suggested by Sydney Caine, who had served before the war as the colony's first Financial Secretary. He pointed out that Young's scheme amounted to the setting up of a diarchy. However, he doubted whether this system was suitable because of the apathy and apprehension of the local people. He also doubted that Young's scheme would provide the right kind of training in self-government. Presumably because of his knowledge that the pre-war administration in Hong Kong had been primarily municipal, Caine thought that Britain was in danger either of establishing a kind of alternative government in the colony or of setting up a sham government in order to avoid introducing an element of greater democracy into the 'real government'. He regarded a change in the Legislative Council as being a more appropriate and more usual way of introducing constitutional advance in British colonies.[19] Whilst Caine was correct in his criticisms of Young's scheme, he failed to provide a viable alternative. If the reforms were to be limited to the Legislative Council, they would not meet the objective of the British government, which was to give the local inhabitants (not just the British subjects) a share in the government.

A third line of thinking was that of T. I. K. Lloyd, successor to Gent as Assistant Under-Secretary supervising Hong Kong affairs. Lloyd studied the subject carefully, discussed it with MacDougall (who was then Colonial Secretary of Hong Kong on leave in England), and came to the conclusion that Mayle's objections, including his point that the introduction of elections would lead to disorder and disturbances, were not valid. He found Young's recommendations on the whole to be 'broadly conceived, well fitted to [the] present conditions in Hong Kong, and workable'. Lloyd stressed that the pledge in May to reform the constitution of Hong Kong could not be reneged on without incurring a serious breach of faith. He thought that the only way to overcome the apathy and allay the apprehension of the local people was to proceed as Young had recommended, which would both show the British determination to stay in Hong Kong and foster an active sense of citizenship in the colony.[20]

The officials at the Colonial Office were in less disagreement over the specific points of Young's proposals. Mayle concerned himself with few of the details because he was basically opposed to any measure of reform. The other officials who commented on the proposals appear to have acquired what Harcourt called the '1946 outlook', although they had reservations about several issues. The equal representation of Chinese and non-Chinese in the municipal council was seen as being open to criticism because the Chinese constituted over ninety-seven per cent of the total population. The officials accepted Young's argument, but insisted that the situation be reviewed later with a view to increasing the Chinese representation. Young's proposal to treat the non-British Europeans and Americans in the colony on the same basis as British subjects in so far as residential requirements for the franchise were concerned was found to be unacceptable because it could be construed by the Chinese as racial discrimination. The allocation of two seats in the municipal council to the trade unions was also considered to be inadequate. Given the preponderance of the working class in the colony, at least three (or even four) seats were considered more appropriate. The recommended property and juror qualifications for the franchise were accepted on condition that they be reduced progressively as education and literacy spread. A review of these qualifications was deemed appropriate after about four years. The rather high minimum age for voters and councillors was also criticized; it was thought that the minimum age for both should be twenty-one. Young's other specific suggestions, including those giving the municipal council full financial autonomy and the fullest scope in its allotted spheres, were accepted.[21]

By the time that the officials in the Colonial Office had resolved their differences as to the general approach towards reform in Hong Kong, it was almost the end of December. It was no longer possible to achieve the original goal of settling and announcing the principles governing the reforms before the end of 1946. It was agreed that there would be no turning back, but that the Governor should be asked to confirm that he still considered the municipal proposal to be appropriate.[22]

Arthur Creech-Jones, who had succeeded Hall as Secretary of State in October, found this turn of events to be 'extraordinary and surprising'. After discussion with his officials, he put forward his own set of proposals for examination and consideration. The main line of reform favoured by Creech-Jones was the reconstitution of the Legislative Council on a liberal basis. Popular elections without communal representation would be introduced, the number of elected unofficial members being equal to that of appointed official members, with an unofficial majority secured by nominated unofficial members. The balance of unofficial and official members would be reviewed after five years. The new Legislative Council would be chaired by a Speaker elected by the Unofficials, but the first Speaker would be nominated by the Governor. Selected unofficial members would also be invited to serve in the 'Government Executive', where they would be associated with particular departments whose affairs they would handle in the Legislative Council. They would also gradually assume some responsibility in the Executive. Creech-Jones also thought that local councils with well-defined functions should be set up in both the rural and urban areas of the colony[23] so that the local people could become acquainted with responsibility by way of carefully regulated entry to the Legislative Council, the central administration, and the local government.[24]

Creech-Jones' main consideration was that Britain be placed 'in a strong position vis-à-vis [its] critics — both [sic] the Pacific Powers, the Chinese and liberals of [the] U.S.A.'. He was also concerned about those in Britain who were critical of and concerned about the question of the future of Hong Kong. Although he recognized the special difficulties in Hong Kong caused by political apathy and the influence of the Kuomintang, he still regarded it as essential that 'a more vigorous policy of development' be pursued as quickly as possible.[25]

Mayle and Caine in the Colonial Office regarded Creech-Jones' proposal as inappropriate. They were opposed to the rejection of communal representation and doubted the wisdom of introducing something like a quasi-ministerial system. The suggestion by Mayle that the Social Services, Communications, Finance, Commercial Relations, and Supplies departments in the Office be consulted was essentially a delaying tactic since the Social Services and Communications departments had nothing to do with the reforms, and the other departments had only a peripheral interest. In any case, the suggestion was rejected by Lloyd, who had just been appointed Permanent Under-Secretary. Young was duly asked at the end of January 1947 to comment on Creech-Jones' alternative proposals.[26]

Young felt strongly that his own reform proposals were more appropriate than those of Creech-Jones. He considered that if the alternative Legislative Council reform were to be adopted, it would be 'unsatisfactory in operation and disappointing' to the responsible local citizens, most of whom favoured a municipal council rather than a widening of the basis of the legislature. Young also stressed the problems associated with placing limits on the franchise if popular elections were to be introduced into the legislature. Although he had no objection to unofficial Legislative Councillors being associated with the work of specific departments, he thought that this association would amount to little if local affairs were to be handled by a municipal council.[27]

The Colonial Office was persuaded by Young's forceful reaffirmation of his earlier proposals to accept his general approach to reform. Mayle, however, still had reservations. He said that 'constitutional advance' in Hong Kong was unnecessary because the existing constitution and administration had attracted and retained a large number of Chinese who clearly preferred the conditions in the colony to those in China. He also stressed that the introduction of constitutional development in the absence of popular demand was 'in effect making a definite and spontaneous move in the direction of the return of Hong Kong to China'.[28] Mayle did not elucidate this last point, which contradicted the views of other officials such as Lloyd and Young. (Both Young and Lloyd — and Gent before Lloyd — considered that the reforms would reassure the local people of the British determination to stay in Hong Kong.) In any event, Mayle's reservations were set aside and the Colonial Office proceeded to prepare a parliamentary statement.

The Foreign Office, the Treasury, and the Prime Minister were consulted at this stage. Opinion in the Foreign Office was divided. One official felt that this attempt to impose Western institutions on an oriental community was unrealistic, and that the 'passion for theoretical equality' between Asians and Europeans was ridiculous. He argued that if Hong Kong were to admit local Chinese to the municipal council, the same rights should be available to British subjects inside China.[29] G. V. Kitson, Head of the China Department and an old China hand, rejected all such arguments. He told Mayle that there was no point in differentiating between the non-British Europeans and the Chinese in Hong Kong. Whilst he admitted his ignorance of the situation in Hong Kong, Kitson was sceptical about whether the trade unions could properly represent the working class; in China itself, the unions were for the most part under Kuomintang control. Kitson also advised the Colonial Office to widen the franchise, which was heavily biased towards the wealthy and the educated. However, the decision was left to the Colonial Office.[30] The Treasury was very critical of the relatively late stage at which it was consulted because, as a result of the war, Hong Kong was at the time under Treasury control. It had no other comment, however, except for its insistence that it be closely consulted and that the Colonial Audit Department (rather than commercial auditors) check the books of the municipal council. The Prime Minister also had no specific comment.[31] The statement which endorsed in principle Young's reform proposals was therefore duly made in the House of Commons on 5 March 1947.

The Colonial Office subsequently examined further those specific points in Young's scheme about which they had doubts. The Colonial Office and Young agreed, after much discussion, that the equal sharing of seats between the Chinese and the non-Chinese in the municipal council should be subject to review in the future. A compromise was also reached on the question of the minimum age for both voters and councillors: it was agreed that it be twenty-five in both cases. With respect to the number of trade union representatives in the proposed council, the Colonial Office agreed with Young that an increase from two to three would be undesirable because it would mean either that one of the other Chinese nominating bodies would need to be dispensed with or that the racial balance would be upset. The differential residence requirements for non-British Chinese and non-British European voters was dropped, however, on the insistence of the Colonial Office. The property qualification for councillors was objected to, by Creech-Jones in particular, because it would preclude some labour leaders from representing their unions. Creech-Jones eventually accepted literacy in English as an alternative to juror qualification in the initial stage.[32]

Young was anxious for his proposals to be publicly approved and published before 17 May 1947, the date of his retirement. He hoped that a firm foundation for the reforms would thereby be laid. Formal approval was delayed beyond that date, however, because of several problems which still had to be settled. Mayle's wish to edit Young's original report to the Secretary of State before it was published (since some of its proposals had not been accepted) was strongly rejected by Young. The Legal Advisor also raised several queries, one of which related to the contradiction inherent in Young's use of the term 'transfer of power' at the same time as he stressed that the authority of the municipal council would be vested in it by an ordinance of the colonial legislature. Since a 'transfer of power' implied that the Legislative Council would surrender certain powers which would then be vested in the municipal council, this would require the authority of Letters Patent or an Order-in-Council. The creation of a municipal council and the transfer of certain powers by an ordinance of the colonial legislature would only amount to a 'delegation of power', which was the usual practice at that time. After his return to England, Young confirmed that it was his considered opinion that anything less than an actual transfer of authority would be regarded as a sham and would cause the local people to lose interest in the reforms. His wish to give the proposed municipal council greater power and autonomy than that enjoyed by municipal councils in Britain confirmed Caine's earlier observation that what Young was proposing to set up was an alternative government in the colony. The Colonial Office clarified this contradiction by substituting the term 'delegation of power', despite Young's recommendations.[33]

After Creech-Jones had consented to his approval and Young's proposals being published before the scheduled arrival of the new Governor in the colony in July, Mayle and his new assistant (A. N. Galsworthy) attempted to introduce a measure which would give the Governor the power to nominate two representatives of the working class if the trade unions were to fail to elect their own representatives. This measure represented a reversal of the earlier attempt by Creech-Jones and Lloyd to increase trade union representation and it was rejected by the other Colonial Office officials.[34]

An additional complication at this stage was the attempt by the British Far Eastern trading community to lobby the Colonial Office to introduce in effect in Hong Kong a replica of the pre-war Shanghai Municipal Council. This effort was undertaken, following Creech-Jones' March statement in the House of Commons, by the China Association, which had close ties with the Hong Kong General Chamber of Commerce. It was not until late July that the Colonial Office convinced the China Association that the proposed municipal council was 'very innocuous' in so far as the interests of the British trading community were concerned. All of the counter-proposals of the China Association were rejected because they were at variance with the government's policy.[35]

The policy makers had deliberated over the constitutional policy for Hong Kong with the future of the colony in mind. The Colonial Office, which had been examining the various possible solutions to the Hong Kong problem since the end of the war, concluded that Hong Kong should not be relinquished.[36] This approach was challenged, however, when the Foreign Office submitted for concurrence in the summer of 1946 a draft joint policy paper on the future of Hong Kong which concluded that Britain should take the initiative in discussing the restoration of Chinese sovereignty over the colony. The Colonial Office responded by preparing an alternative paper. Lloyd also attempted to enlist the support of the Treasury against the Foreign Office by stressing that any initiative along the lines advocated by the Foreign Office would drive investors away from the colony and reverse its economic recovery, forcing the Treasury to pay for the consequences.[37] Creech-Jones also stressed that the policies for constitutional reform and the future status of the colony were complementary.[38] Young, who strongly opposed the Foreign Office's suggestion, went so far as to use his reform plan to dissuade J. L. Stuart, the American Ambassador to China, from putting pressure on Britain to give up Hong Kong. Young told Stuart that he wished to set up a model municipal government in Hong Kong 'as a contribution to Chinese progress' but that this would require at least thirty years.[39] He was attempting to secure thirty years of British rule in the colony in order to make it possible for a new generation of citizens who would be loyal to British Hong Kong to participate in the government before the problem of the colony's political status was tackled.

When the Colonial Office originally authorized Young to make his 1 May 1946 statement, it had recognized that Hong Kong's predominantly Chinese character would require special attention. However, it had not realized the full dimension and complexity of the related problems of political apathy, corruption, the danger of Kuomintang infiltration, and the sensitivity of the residents over the uncertainty of the future of the colony. When the full dimension of these problems was appreciated, the only official who recommended that the Colonial Office depart from its original intention was Mayle, a very conservative man.[40] However, Mayle's anti-reform stance caused little delay. Creech-Jones' preference for an alternative approach also had little impact, having no real support, and he did not insist that it be adopted. His idea of introducing a quasi-ministerial system had in fact been accepted by Young, but it was not followed up by the Colonial Office.

On 24 July 1947, one day prior to the arrival of the new Governor, the Hong Kong government published Creech-Jones' approval in principle of Young's reform proposals. The proposed modifications to the Legislative Council were accepted without comment. The various changes concerning the municipal council scheme, which had been agreed by the Colonial Office and Young between March and July, were made public. The question of whether the municipal council should be subject to any form of external control was reserved for further examination, however, because of the potentially important legal and financial problems. The Hong Kong government was also given the express authority to proceed with detailed preparations in order to give effect to the Young Plan. Hong Kong was, finally, on the road to constitutional advance.[41]

The Restoration of Civil Order

Immediately following the ceremonial restoration of civil government on 1 May 1946, Young convened a special meeting of the Legislative Council in order to make constitutional provision for the continuance of various measures taken by Harcourt's Military Administration. Mindful of the pending constitutional reform, Young also made a number of changes when he re-established the Executive, Legislative, and Urban Councils. First, he limited the period of appointment of unofficial members of all councils to one year. Secondly, when the Urban Council was re-established, he did not reinstate the pre-war practice of election of two of the unofficial Urban Councillors, pending a decision on the municipal reform proposal. The Urban Council therefore resumed its duties as the government agency for sanitary and other related matters with five official and six nominated unofficial members.[42]

More importantly, Young fully realized the need for a '1946 outlook', and he took practical steps to mark the beginning of a new era. He reconstituted the Executive Council (which before the war had comprised seven official members, two expatriate British unofficial members, and one Chinese-British unofficial member) by adding an additional Chinese-British unofficial member, thus giving equal weight to both the Chinese-British and the expatriate British communities. He also repealed the pre-war legislation which had restricted the Chinese from residing in the Peak district. Young added substance to this gesture not only by reiterating the policy of localization, but by appointing, for the first time in the history of Hong Kong, an ethnic Chinese to the Cadet Service.[43]

In order to meet the special requirements of post-war reconstruction and rehabilitation, Young also created new temporary departments in addition to restoring the old ones. The biggest problem in the initial stage of this process was the temporary shortage of senior officials, many of whom had not recovered sufficiently from internment during the war to return to their duties. Young overcame this difficulty by retaining a number of military officers of Harcourt's Administration on one-year contracts. This measure filled an important gap without committing the government to a permanent increase in its establishment. Nevertheless, in order to continue Herklots' work on the rehabilitation and development of local fishing and agriculture, Young defied the tradition of reserving the top posts for Cadets (and for senior officers transferred from outside the colony) by appointing Herklots, formerly a university lecturer, as head of a new Secretariat for Development.[44]

Young also recognized the importance and delicacy of the Sino-Hong Kong relationship in the new era. He considered it essential that the Hong Kong government take a more positive approach to Chinese affairs. For this purpose, he recommended the appointment of a Political Adviser who would be a specialist on Chinese affairs. He suggested that the first officer be seconded from the Foreign Service because most of the senior officials in the government were out of touch with events in China. The first appointment was not made, however, until after Young's retirement because the Foreign Service was also short of China experts.[45]

The restoration of civil rule did not in itself solve any of Hong Kong's economic problems. The war and the Japanese occupation had reduced Hong Kong to financial dependence upon the Treasury. In a bid to end this dependence and to secure a source of steady income with which to modernize the government, Young advocated the introduction of income tax. Although such a move was strongly opposed by both the Chinese and the expatriate British communities, the need to balance the budget left Young with little choice. The income tax legislation, which was based on the War Revenue Ordinance, 1941, was pushed through despite the determined opposition of three unofficial members of the Legislative Council. In introducing this measure, Young made it clear that he saw the introduction of income tax, the reform of the Legislative Council, and the establishment of a municipal council as a package. Both direct taxation and representation were to be part of the new deal.[46]

The restoration of civil government helped to rehabilitate trade. It provided the legal framework within which commercial transactions were regulated; it ensured the security of goods and investments; and it maintained general stability. The environment was thereby provided in which traders could resume their operations and tap the available sources of supply in order to meet the requirements of a market starved of basic essentials and consumer goods. It also meant that a secure base was established from which Hong Kong traders could operate in the Far East where political instability was the order of the day. By 1946, the value of total visible trade (calculated in Hong Kong dollars) was $1.7 billion, well above the 1940 figure of $1.375 billion. (However, these figures do not reflect the change in the value of the dollar over the period.) It has been estimated that Hong Kong handled in 1946 approximately fifty–sixty per cent of the volume of pre-war trade. A comparison of the 1946 and 1947 figures reveals that the total value of visible trade in the first five months of 1947 was $987.5 million and the value for the whole year was $2.767 billion — an increase of just less than $1 billion over the 1946 total.[47]

In spite of the impressive trade recovery, the rehabilitation of the local economy was not without difficulties. The main problems in May 1946 were the supply of rice and fuel. The shortage of rice was so serious that the daily ration was reduced to three ounces per person per day (compared with twelve ounces in December 1945).[48] The supply problem had only just begun to ease when the shortage of housing and the high cost of living became of major concern. The living conditions of the working people were so harsh that in the first five months of 1947 alone, twenty-one industrial actions occurred, all but three of which were economically motivated. (The three exceptions were aimed generally at improving working conditions.) Although the retail price index began to fall towards the end of Young's governorship, the standard of living of the working people did not improve at that time to any degree. Between June and the end of 1947, thirty-four industrial actions took place, thirty-three of which were clearly economically motivated. It is noteworthy that none of the industrial actions during the year were politically motivated.[49]

It is significant that neither the rapid recovery of trade nor the harsh living conditions experienced by the working people were conducive to the introduction of Young's reforms. The traders' concern with the recovery of their businesses and the preoccupation of the workers with their struggle to earn a living contributed to the general lack of enthusiasm with which Young's announcement of reform was greeted by the local inhabitants.[50]

The Challenge of the Chinese Political Parties: I

Young's return to Hong Kong on 1 May 1946 coincided with the high tide of Kuomintang activities in the colony. This activity represented for Young a special problem in his attempt to introduce constitutional reform. Based on the observations of the Acting Secretary for Chinese Affairs, Thomas Megarry, Young concluded that the Kuomintang had established extensive influence and control over the local Chinese community. He believed the claim of the Kuomintang that it had control over 120 organizations, including vernacular newspapers, schools, and labour and commercial organizations. He thought that the aim of the Kuomintang was to build an 'imperium in imperio', and that its immediate objective was to assume complete control over the vernacular press and the labour organizations. He also accepted as evidence of the Kuomintang's success its take-over of the hitherto independent *Wah kiu yat pao*, the largest circulating Chinese-language newspaper; its apparently burgeoning influence over the Chinese Chamber of Commerce; and (with the exception of the supporters of the Chinese Communist Party and other anti-Kuomintang Chinese political parties) the reluctance of the local Chinese to antagonize the Kuomintang. The zeal with which the local Kuomintang celebrated the Chinese national day was also interpreted as an indication of the indifference of the Kuomintang to Hong Kong's British identity.[51]

Young's assessment of Kuomintang influence in the local Chinese community caused considerable concern when it was presented to the Colonial Office. The Kuomintang had a high profile, and it was prepared to air its views on all matters concerning the relations between Hong Kong and China. This higher level of activity compared with the pre-war years did not necessarily mean that the Party was successful in forming what Megarry called an 'imperium in imperio'. The Kuomintang had been banned before the war and it had had to proceed warily until its existence became legal and open after the war. As the nationalist party of China, it was expected to make its presence in the colony known. However, the true extent of its influence and control over the Chinese community was unknown.

The attempt by the Kuomintang to take over the *Wah kiu yat pao* was (in Megarry's opinion) a prelude to a take-over of the Hong Kong Chinese-language press. The Canton authority arbitrarily pronounced the rightful owner of the newspaper, Sum Wai-yau, a traitor. The local Kuomintang then launched a public campaign against the newspaper, with the *National Times* (the *Kuo min jih pao*), a Hong Kong-based Kuomintang-published newspaper, as its spearhead.[52] Before this campaign produced the desired result, however, the *National Times* was suspended for a month by the Governor-in-Council on a charge of inciting violence. The Canton government then asked for Sum's extradition, which was refused. Sum eventually agreed to turn over a majority holding of the newspaper to a Kuomintang general in exchange for the removal of his name from the traitors' list. The incident, which lasted for over six months and ended in the nominal take-over of the newspaper by the Kuomintang, was in fact only a hollow victory. The newspaper was not turned into a Kuomintang propaganda machine and its editorial policy remained independent. More importantly, and contrary to Megarry's opinion, the *Wah kiu yat pao* was not used by the Kuomintang as a spearhead for a take-over of the local vernacular press since no subsequent take-over attempts were made.[53]

Megarry also misunderstood the significance of the burgeoning efforts of the Chinese Chamber of Commerce to curry favour with the Kuomintang. The Chamber's efforts in this direction were merely good business practice, given that most Hong Kong businessmen had trading interests in China, that the Kuomintang was the governing Party in China, that the Party was by then utterly corrupt, and that (as was not the case in the pre-war days) it was officially represented in Hong Kong. A more accurate yardstick against which to measure the local traders' support of the Party was the extent of their contribution to Party funds. The Kuomintang's appeal for contributions (either for the general purpose of financing its party apparatus or for the specific purpose of constructing a Sun Yat-sen Memorial Hall) in fact met with little response.[54]

Megarry also wrongly assessed the control which the Kuomintang exercised over labour. He described the Kuomintang as gaining ground and the Chinese Seamen's Union, which had its headquarters in Shanghai and a recently established branch in the colony, as 'a powerful tool of the Kuomintang'. The Union failed, however, to weaken the position of the existing left-oriented Hong Kong Seamen's Union, which continued to function as before. Although the biggest strike during this period, which involved more than 10,000 workers, was initiated by the pro-Kuomintang Chinese Engineers' Institute in August 1947, the strikers were not politically motivated; in fact, they turned down the Kuomintang's offer of financial support. The other local unions which supported the Kuomintang were generally both weak and politically inactive. (None of the fifty-five industrial actions in 1947 were politically motivated.[55])

The support which the Kuomintang appeared to enjoy in Hong Kong was often illusory. The Chinese patriots in Hong Kong who identified themselves with China did not necessarily also support the Kuomintang. This distinction was subtle, however, and difficult to detect. The Kuomintang acted as if it were the Chinese government and its emblem was very similar to the national flag of the Republic of China. Indeed, the Kuomintang emblem — a white sun in a blue sky — is incorporated as an integral part of the design of the national flag: a white sun in a blue sky, set against a background of red. The lyrics of the national anthem of the Republic of China even begin with a direct reference to the Kuomintang: 'San-min-chu-i wu tang so tsung' ('The Three People's Principles are the Party's motto'). Any Chinese nationalistic outburst at which the Chinese national flag was shown and the Chinese national anthem was sung could therefore be interpreted as support of the Kuomintang unless the anti-Kuomintang stand of the participants was expressly declared, which was not always the case.

The securing of official recognition was the greatest achievement of the Kuomintang in Hong Kong after the war. This recognition was due less to any strengthening of the Kuomintang in the colony than to the change in the international situation and China's rise to Great Power status. The Kuomintang's finest moment was the first openly held annual general meeting of its Hong Kong and Macau Branch in July 1947. By its own assessment, the local Kuomintang had made little progress since the end of the war. Although it had planned to recruit 50,000 members, it managed to gain only between 8,000 and 10,000. The organization of the Party in the colony was also very loose: the senior and junior ranks maintained contact generally only through correspondence, and party officials of the same rank seldom co-ordinated their activities. Party officials tended to be passive, self-seeking, and bureaucratic in their outlook, and there was little sense of comradeship. Propaganda work was restricted to publishing, which had only a limited effect on the intellectuals. The Party consequently remained 'top-heavy', with no grass-roots support.[56]

Li Ta-chao, head of the local Kuomintang, defined the aims of the Hong Kong Branch as the destruction of the Communist bandits, the achievement of peace, the unification of China, and protection of the fruits of the victory in the war — aims which contrasted sharply with those attributed to the Kuomintang by Megarry. The means by which these aims were to be realized were, first, the building up of a broad basis, the institution of a system of promotion according to merit, a reduction of bureaucracy, and general preparation for its participation in the new democratic government about to be set up in China. Secondly, expansion of the Kuomintang's influence over the young people was deemed to be particularly important because of the Communists' efforts (similarly unsuccessful) in this direction. Held to be of importance, also, was the development of a sense of comradeship and responsibility amongst the local Party members.[57] Although the Kuomintang would have been unlikely not to welcome an extension of its influence over the local Chinese, there is nothing in the available Kuomintang materials to suggest that its aim was to actively build up an 'imperium in imperio' in Hong Kong.

Megarry's failure to understand the true situation was apparent also in his interpretation of the strong Kuomintang bias with which the Party in Hong Kong celebrated the Chinese National Day. Whilst he interpreted the celebrations as a belittlement on the part of the Kuomintang of Hong Kong's status as a British colony, such an interpretation is unreasonable under the circumstances. The 1911 Revolution in China was the Kuomintang's revolution, and the Double-Tenth was not merely the Chinese National Day but also the anniversary of its revolution. A strong Kuomintang bias to the celebrations was therefore to be expected. Hong Kong's British identity was in fact acknowledged, and the local Kuomintang emphasized in its commemorative propaganda material the need for China to develop further its relations with Britain and the United States. If the local Kuomintang had not been well-disposed towards British authority in Hong Kong, it would have demanded its retrocession to China.[58]

The most important indication of the Kuomintang's policy towards Hong Kong was the attitude of Chiang Kai-shek. Shortly after Japan announced its surrender in August 1945, and in the course of the Sino-British dispute over the liberation of Hong Kong, Chiang publicly declared his view:

I wish to state here that the present status of Hong Kong is regulated by a treaty signed by China and Great Britain. Changes in future will be introduced only through friendly negotiations between the two countries. Our foreign policy is to honour treaties, rely upon law and seek rational readjustments when the requirements of time and actual conditions demand such readjustments .... China will settle this last issue [i.e., the lease for the New Territories] through diplomatic talks between the two countries.[59]

Chiang's attitude had not changed since the end of the war. In a meeting in May 1947 with the British Ambassador, Chiang went further and said that he agreed to the British taking a neutral attitude towards fugitive Chinese political leaders (that is, his own political opponents) in Hong Kong. There is no doubt, therefore, that Chiang respected Hong Kong's British identity at that time.[60]

Of particular importance in the context of this study was the attitude of the Kuomintang towards the reform proposals. Shortly after Young made his radio broadcast of 28 August 1946, Li told the press that he was strongly in favour of self-government for the people of Hong Kong, and that the Kuomintang would do its best to assist the Hong Kong government to implement its plan. He also urged the local Chinese to play an active role in the reforms, for their own (rather than for China's or the Party's) benefit. When Young's proposals were publicly approved, the Kuomintang gave its endorsement in terms which made the editorials of the independent vernacular newspapers sound parochial. Although the Kuomintang objected to the equal (yet unfair) division of seats between the Chinese and the non-Chinese in the proposed municipal council, it accepted it because Hong Kong was British.[61] At a much higher level, Sun Fo (Vice-President of China and a senior Kuomintang leader) also publicly advised the local Chinese, when he visited the colony in August 1947, to support the introduction of municipal reform. He welcomed the Young Plan, which he considered an application to Hong Kong of the new British colonial policy which had already been introduced in India and Burma.[62]

Although the Kuomintang had little real success in extending its control over the local Chinese, and in spite of its accommodating attitude towards the colony, it nevertheless posed a threat to the Hong Kong government. It was organized, and it could appeal to the nationalist sentiment of the local Chinese in the event of a show-down with the British. The likelihood of such an event was remote, however. The principal tasks of the local Kuomintang were to counter Chinese Communist propaganda and to rally local Chinese support for Chiang Kai-shek, not to subvert the Hong Kong government.[63]

Young's lack of knowledge of the Chinese or of Chinese politics was evident in his failure to understand the real position of the Kuomintang in the colony. The disturbing state of affairs as described by Megarry continued unchanged, in Young's view, throughout his one year in office. (In reality, however, the situation was never as disturbing as Megarry had indicated.) After careful consideration, Young concluded that the only solution was to expel the Kuomintang from the colony. With this solution in mind, he instructed the Defence Security Officer and the Director of the Police Special Branch to keep the Kuomintang under close surveillance and to make the necessary preparations for its immediate expulsion if and when a suitable excuse could be found. By April 1947, the Police had prepared a 'reasonably complete dossier' on the leading Kuomintang infiltrators and it was capable of expelling a large number of them at short notice. This state of preparedness indicates that, however serious the perceived Kuomintang threat, Young considered it to be manageable. More significantly, Young still considered his reform proposals to be appropriate.[64]

The Chinese Communist Party was also perceived as a potential threat to the Hong Kong government. As was the case with the Kuomintang, Communist activities in the colony were aimed at aiding its struggle for power in China proper. The Chinese Communists in Hong Kong had three major goals: to promote the United Front against the Kuomintang, and hence to rally support for its own cause; to support guerrilla activities in Kwangtung province; and to use the colony as a contact point with

the outside world. For these purposes, and because Hong Kong
was its only safe sanctuary in the region, the Communists surrep-
titiously located their South China Bureau in the colony. The
Bureau was reportedly headed by an alternate member of the
Central Committee of the Chinese Communist Party, Liao
Cheng-chi, and the Head of the Kwangtung guerrilla forces, Fang
Fang. The Communists also maintained an unofficial (yet visible)
presence under the guise of the Hsin Hua News Agency (that is,
the New China News Agency). Chiao Kuan-hua (alias Chiao Mu)
was the director of the Agency and a *de facto* Chinese Communist representative. The strength of the Communist organization in
the colony was unknown, but it was reliably estimated at about
5,000 supporters.[65]

Although the Communist Party used Hong Kong as its major
base for operations in south China, it was scrupulous in its activities. The Communists knew their position and the rules of the
game. The Party had been outlawed as an organization, but its
supporters were at liberty as long as they did not break the laws
of the colony and were not openly hostile to the government of
China, a friendly government in so far as the British were concerned. The Police Commissioner was subsequently given no
cause to make a complaint about them. The Communists also
avoided making any hostile gesture towards the British. Indeed,
David MacDougall, Hong Kong's Colonial Secretary, found the
Hsin Hua News Agency to be 'the only [news agency] which
adopted a dispassionate, and in fact often friendly, attitude towards the Colonial Administration'. This view was shared by
senior officials in the British Embassy in Nanking, who found the
Chinese Communists so well behaved that when the Hsin Hua
News Agency applied in late 1946 for permission to operate a
two-way radio between Yenan (the capital of the Communist
zone in China) and Hong Kong, they could find no justification
for turning down the application.[66]

The local Communists also supported Young's reform proposals. While pointing out politely that the provision for working-
class representation could be improved, they hailed the reforms
as an important move towards democracy in the colony and a
contribution to democratic development in China. They gave no
indication of any intention to infiltrate the proposed municipality.
They were also lavish in their praise of the Labour government,
describing Creech-Jones as an 'enlightened statesman'.[67] It is

worthy of note that Young completely ignored the Communists in his public and private dispatches concerning the reforms — an indication that he thought them to be unimportant.

Traders, Councillors, and Reform

The articulate non-Chinese community in Hong Kong was generally sceptical of Young's idea of reform. The strongest opposition was that voiced by two expatriate Britons who rejected, in racially contemptuous language, any form of self-government for the local inhabitants. They said that the Chinese, Eurasians, Portuguese, and Indians in Hong Kong had only the interests of their own countries in mind. If these people were to take part in self-government, they would discriminate against other nationalities in the name of the British government in Hong Kong, resulting in 'an era of unparalleled corruption, discrimination, injustice and strife'. Such people should be left to 'rot' and 'stink'.[68] As Harcourt had predicted, some of the old-timers who had returned to the colony failed to realize the need for a new post-war outlook, and the tone of their objections to reform was reminiscent of the pre-war days. The English-language press generally recognized that some form of gradual reform was necessary, though it doubted the wisdom of Young's proposals. It objected to the idea of a municipal council, either because such a council was thought to be superfluous and of little importance or because it would mean embarking upon an experiment in difficult times. The reform of the Legislative Council was deemed a more appropriate alternative than direct representation, which would involve the introduction of income tax. One proposal was that indirectly elected members should replace nominated members in the Legislative Council.[69]

The British trading community in Hong Kong, which was heavily represented in the colonial legislature, also took exception to many of Young's proposals. Following Young's invitation for views and comments, the Hong Kong General Chamber of Commerce appointed a special committee under its Chairman, R. D. Gillespie of Imperial Chemical Industries, to look into the issues involved. Like Young, the committee favoured the establishment of a municipal council with some corresponding adjustment in the constitution of the Legislative Council. It also considered it important that the council be given wide powers so that it would attract responsible people to participate in it. With respect to the other proposals, however, the committee was in disagreement with Young. It preferred a small council of sixteen members, consisting of ten non-Chinese and six Chinese unofficial members, producing a clear non-Chinese majority. The non-Chinese councillors would be elected by secret ballot and the Chinese councillors by electoral colleges. Membership of the council would be limited to British subjects. In essence, except for the last provision, the committee was proposing the resurrection in post-war Hong Kong of the pre-war Shanghai Municipal Council, even to the extent of its size.[70]

The China Association — the representative body of British Far Eastern trading interests in London and a sister organization of the Hong Kong General Chamber of Commerce — shared the views of the Chamber, having kept in very close contact with it on the subject of constitutional reform. When it became known in the spring of 1947 that Young had produced a scheme materially different from the Chamber's proposals, the China Association promptly lobbied Whitehall for its (and the Chamber's) proposals. It approached both Creech-Jones and the head of the Hong Kong Department in the Colonial Office, though with little real success. The Association later admitted that it had 'allowed [its] own mind to run somewhat too much on [its] conception of a Municipal Council as derived from experience ... in Shanghai'.[71]

The British traders conceded that it was necessary to involve the Chinese in the management of Hong Kong, but they were adamant that the power of veto must remain in their own hands. Whilst they recognized the need for a new post-war system, they did not share the views of Creech-Jones or the British government. Young's pledge to give the local inhabitants of Hong Kong a share in the management of their own affairs was of little importance to the British traders.

The views expressed by the Chamber and the China Association represented more than the opinions of the traders alone; they also represented the majority view of the expatriate British Legislative Councillors. The Chamber's special committee to study the reforms was constituted by Gillespie (who was concurrently the Chamber's representative in the Legislative Council and Chairman of the Hong Kong Branch of the China Association), D. F. Landale (who sat on both the Executive and Legislative Councils), C. C. Roberts (at times Legislative Councillor),

J. R. Jones (a former Secretary of the pre-war Shanghai Municipal Council), and G. Miskin. Two of the three serving expatriate
British Legislative Councillors[72] were therefore actively involved
in drafting the proposals of the Chamber. Both expatriate British
unofficial Executive Councillors (Arthur Morse and Landale)
were also at that time leading members of the Chamber and the
China Association. In view of this, it is remarkable that the
traders should have decided to lobby Whitehall rather than Government House.

The opinions of the unofficial Chinese Executive and Legislative Councillors were conspicuous by their absence. The three
Councillors seldom spoke on the subject, either in the Legislative
Council or to the press. M. K. Lo, a member of both Councils,
did say in the Legislative Council that one of his 'visions and
dreams [is that the Hong Kong people] ... aided by sound
education, will assume progressively and in ever-increasing measure, the responsibility of self-government'. On that occasion,
however, Lo did not elaborate on the kind of self-government he
was dreaming about or when he expected his dream to
materialize.[73] Judging by his comments on other occasions, Lo
presumably wanted opportunities for the locally recruited civil
servants to rise to the top of the administration.[74] The only other
Chinese Legislative Councillor to openly refer in the Legislative
Council to the reforms was S. N. Chau, who said that taxation
should be linked to representation. He did not elucidate this
point, however, but merely opposed the introduction of income
tax and, by implication, the reform.[75] Lo explained the silence of
the Unofficials as an attempt not to influence public opinion in
the light of the Governor's invitation to the public to make their
views known directly to him.[76] Such rationalization is hardly
convincing. Had they wished to do so, the Unofficials could have
encouraged the general public to speak out on the subject without directing them to any particular course of action. Whilst they
did not like Young's approach, for reasons of their own the
Unofficials simply chose not to oppose Young.[77]

The Attitude of the Local Chinese

When Young restored civil government in Hong Kong, the expectations of the local Chinese community were generally very
modest. It was hoped that Young would continue what Harcourt

had begun and solve the pressing problems of supply and housing, maintain law and order, promote economic recovery, improve Sino-British relations, and generally take into account the
opinions of the people. The local Chinese expected from the
Governor more or less what they would have expected from a
father. In their estimation, a governor was a *fu-mu-kuan* — a
'father and mother official' — a far-sighted and capable mandarin
who knew the needs of his people and who would care for them
like a father would care for his sons.[78]

Young's announcement of the British government's intention
to reform the government was welcomed as a wise decision. The
announcement was fairly widely covered but seldom commented
on in the vernacular newspapers. It was felt that the Governor
had decided wisely and with foresight, and that the people should
act worthily and make the reform a success. Nevertheless,
although Young's subsequent appeal to various public bodies for
their opinions was heartily applauded because it showed his concern for the local people, there was very little public comment.
The dearth of comment may be attributed to the belief that the
Governor and Whitehall had already prepared a blueprint for
reform in London. It may also have been due to the fact that
the local community leaders had admitted that they were not experts on constitutional and administrative matters. Under such
circumstances, it may have seemed pointless for them to put
forward their own proposals, the best approach being for the
Governor to make public the draft proposals which would then
be subjected to public discussion.[79]

This state of affairs began to change slightly after Young outlined his reform proposals in his August radio broadcast. As a
direct response to the Governor's fresh appeal for local Chinese
opinions, the *Hsin sheng wan pao* launched a campaign to elicit
public opinion, culminating in a seminar on 27 September to
discuss the issues involved. The *Hsin sheng wan pao*'s claim that
it had received views representing several thousand Chinese inhabitants in the four weeks subsequent to Young's broadcast was
not the beginning of a deluge of opinion, but it compared favourably with an opinion poll published in June in two of the local
English-language newspapers which had received only a hundred
replies. Young's broadcast enabled many interested Chinese to
develop their own views, while at the same time confirming that
the Governor genuinely wished to hear from the public.[80]

The expressed opinions of the local Chinese included a wide range of views, but Young's municipal approach was generally accepted. With respect to the scope of the council, it was thought that it should be as wide as possible but that defence, foreign policy, and the Police should be reserved to the central government. It was also agreed that the council should consist of appointed, indirectly elected, and directly elected members. Public opinion favoured a Chinese majority in the allocation of seats because the Chinese constituted over ninety-eight per cent of the total population, but it was also demonstrated that a token majority for the Chinese in the council would be acceptable — an indication that 'face' was the main consideration. It was agreed that the election of councillors should be based upon some form of residential, age, property, educational, or professional qualifications, and that similar, though less restrictive, requirements should apply for voters. Some other aspects of Young's proposals were also discussed by various organizations, their primary concern being simply that there be no racial discrimination in the new council against the Chinese.[81]

Any assessment of the response of the local Chinese to the idea of reform must recognize its traditional character: important policy ought to originate at the top, after careful study by the experts. The local leaders would support it if it was acceptable and applaud it if it was commendable; anything further was unnecessary unless the policy was unacceptable. The vast majority of the illiterate and semi-illiterate considered public affairs to be the domain of the government and of the educated; their main concern was to make ends meet. This mentality was therefore a crucial factor in the response of the local people to Young's announcement of reform.[82]

Another important consideration was the prevailing economic climate. The working people, who included clerks and many educated Chinese (hence, potential voters), were still struggling to earn a living. Such people had little time to study the matter of reform and to submit their views. Nevertheless, there were a certain number of Western-educated Chinese who did not share this mentality and who could examine the issue in depth at their leisure. Their opinions were not often made available to the Governor, however, because of their cynicism about the government. One such Chinese, for example, studied Young's August broadcast closely. He preferred an alternative approach, but sent his considered views (five pages in length) to the Fabian Colonial Bureau rather than to Young.[83]

On the whole, there was no vehement demand during Young's governorship for a complete reform of the government in Hong Kong. The overwhelming majority of the local inhabitants expressed no views at all on the issue. The articulate Chinese generally supported Young's proposals, though not without some reservations on certain matters of detail. By the time of Young's retirement, it was accepted that there would be some move towards constitutional advance; it remained only for Whitehall and the new Governor to make the formal announcement.

4 Interlude

ALTHOUGH Secretary of State Creech-Jones publicly announced his approval of Young's reform proposals (the Young Plan) on 24 July 1947, no significant step to implement the Plan was taken for one and a half years. Then, towards the end of 1948, Young's successor as Governor, Sir Alexander Grantham, intimated that the unofficial members of the Legislative Council intended to put forward an alternative to the Young Plan. A policy review at the time resulted in a decision which effectively meant that the Plan was put aside.

This chapter examines the events leading to this unexpected turn of events and attempts to identify the principal cause of the long delay in the implementation of the reforms.

When Grantham arrived as the new Governor of the colony on 25 July 1947, one day after the approval of the Young Plan was made public, he indicated that because he thought there was general agreement as to what should be done he had come with no new policy, no new mandate, and no new brief. It merely remained for the Plan to be implemented. He assured the Hong Kong people that Hong Kong was about to take 'a great step forward on the democratic path by the establishment of a municipality and ... the creation of an Unofficial majority in the Legislative Council'.[1] He further stressed that 'it would be a very great mistake ... if there were to be a new policy, because that would mean the stopping of everything that was being done and starting all over again. It would at best mean a long delay'.[2] The new Governor clearly intended to put his predecessor's approved policies into practice without undue delay.

Despite his assurances to the Hong Kong people that reform was in the pipeline, Grantham was not at all enthusiastic about the Young Plan. In fact, he had had reservations about it even before he became Governor. When he had been briefed on the subject as Governor-designate in London, he had made it clear that he disagreed with Young's idea that the proposed municipality should be exempted from external control. He preferred that the Governor-in-Council have special powers in order to guard against any failure in its duties on the part of the municipality.[3]

More fundamentally, having come to know the Chinese and the colony as a junior Cadet between 1922 and 1935, Grantham felt that the Young Plan was not suitable for Hong Kong. In his view, 'a benevolent autocracy' was the appropriate form of government.[4]

Grantham appears to have begun to deal with the Young Plan in mid-August 1947. He received a deputation from the Hong Kong General Chamber of Commerce which merely reiterated many of its arguments.[5] He also authorized the printing of three draft bills which were necessary for the establishment of a municipal council: the Municipal Council Ordinance, the Municipal Electors Ordinance, and the Corrupt and Illegal Practices Ordinance. These draft bills, 'which need[ed] alteration only on points of detail' before they could be introduced, had been drafted by Hazlerigg prior to Young's retirement.[6] In early October, Grantham appointed a class two Cadet, J. H. B. Lee, as Assistant Colonial Secretary (Special Duty) to take over the duties of Hazlerigg, Young's Special Adviser, who had retired and left Hong Kong in May. Unlike Hazlerigg, however, Lee was not given a seat in the Executive Council (or, for that matter, in the Legislative Council). At about the same time, Grantham also circulated the draft bills to members of his Executive Council. In mid-November, he sent the draft bills to the Colonial Office, saying that they would soon be discussed in the Executive Council. Up until this time, there was no indication that he intended to reverse Young's policy on constitutional reform.[7]

Notwithstanding Grantham's announcement to the Colonial Office in mid-November, there is no record in the Executive Council minute for the first year of Grantham's governorship that the draft bills were discussed in the Council.[8] Subsequent events, and the fact that Grantham appears to have changed his mind after mid-November with regard to the Young Plan, indicate however that he must have discussed the draft bills with the unofficial members of the Executive Council. Since Grantham had very close relations with the Unofficials, he may have discussed the matter with them privately and then decided not to put the bills to the Council officially. Alternatively, the Executive Council may have discussed the issue but, for reasons of their own, Grantham and/or the Unofficials may have decided that the discussion should not be recorded in the Executive Council minute — not an unusual practice. It would not be surprising for

Grantham to decide to proceed slowly if, having reservations
himself about the Plan, the Unofficials also indicated that they
did not view it favourably. Instead of proceeding with the practical preparations for the reforms, as he had been asked to do
in the Secretary of State's dispatch approving the Young Plan,
Grantham therefore waited for further instructions from the Colonial Office without notifying it of his decision not to act.

In the meantime, Mayle and his assistants in the Colonial
Office had been examining the question of external control over
the municipal council — a matter reserved for later decision by
the Secretary of State. In contrast to the pace with which the
whole matter of reform had been considered before July, this
issue was handled in a leisurely manner. Such an attitude is
understandable in the light of Mayle's continued scepticism about
the Young Plan. Nevertheless, Mayle and A. N. Galsworthy, one
of his assistants, thought that external control was necessary.
Strict financial control over the municipality was deemed inevitable in view of the Treasury's control over Hong Kong, and the
confirmation or disallowance by the Legislative Council of municipal by-laws was also considered to be essential. Indeed, they
thought that, if the Governor found it politically feasible, the
municipality should be required to act in general conformity to
the policy of the colonial government. They also considered it
desirable that the Governor-in-Council should have special reserve powers if the municipal council were to default on any of its
duties. All of these provisions directly contradicted Young's original proposals and the wishes of Creech-Jones, and they were
reserved for further consideration precisely because they were
potentially controversial. G. F. Seel, the new Under-Secretary
supervising Hong Kong affairs, nevertheless sanctioned the provisions without reference to Creech-Jones. Further delay was
caused by another of Mayle's assistants, W. I. J. Wallace, who
was preoccupied with work concerning the report of the Hong
Kong Civil Service Salaries Commission. It was not until late
October, four months after Creech-Jones had approved the reforms, that a draft dispatch setting out the Secretary of State's
further observations on external control was finally prepared.
This dispatch was sent to the Treasury for concurrence, and to
Grantham for his information and comment.[9]

The Treasury found the problem to be intractable. It rightly
pointed out that the Young Plan was an attempt to divide power

and functions between the municipal council and the colonial government — an operation which had not been attempted elsewhere in the British Empire. With respect to the question of financial control, Treasury officials realized, in principle, that the municipality should be given a proper sense of financial responsibility. However, in practice, in the light of Treasury control over the Hong Kong government, the Treasury deemed it invidious to exclude the municipality from its control. More specifically, it meant that the municipality should be subject to strict financial scrutiny by both the colonial government and Whitehall. The Treasury also made it clear, however, that it would be willing to review the arrangement once Hong Kong had ceased to receive Treasury subvention.[10]

A total of five months had been required for the Treasury to reply to the Colonial Office's October 1947 invitation for views and/or concurrence. This delay should not have prevented the necessary practical preparations being made in Hong Kong, however, because Creech-Jones had already expressly authorized the Governor to proceed. In any case, the Secretary of State's observations on external control were amended further in the light of the Treasury comment and were formally sent to Hong Kong before the end of March 1948.[11]

The Hong Kong government had meanwhile, apparently, taken no steps to prepare for the implementation of the Young Plan. The three draft bills which had been sent to the Colonial Office in November 1947 had not even been amended in the light of the Secretary of State's comments in his July 1947 dispatch approving the Young Plan. It is unclear just what Lee did during his appointment as Assistant Colonial Secretary responsible for the reforms, which in any case lasted for only five months before he was reassigned in February 1948 to become Postmaster-General. The government had made no public reference to the reforms until the Governor made a report in his annual address to the Legislative Council on 19 March 1948; he regretted that progress had been slow.

Grantham attributed the slow progress in the implementation of the reforms to the fact that the amount of detailed work to be done had been underestimated. He referred to the need to set up the technical and legal framework, including matters such as the delineation of voting wards, the compilation of voting registers, the drafting and scrutinizing of the necessary ordinances, the recruitment of an experienced town clerk and a special officer to oversee the proper launching of the municipal council, and the finding of the physical site to house the council. He also stressed that the necessary legislation would be laid before the Legislative Council as soon as it had been commented on by the Secretary of State.[12] Grantham did not mention, however, the measures which his administration had taken to overcome these legal and technical difficulties.

Notwithstanding the difficulties Grantham referred to, his explanation for the delay in implementing the Young Plan is unconvincing. He failed to explain, for example, why these practical difficulties had not been overcome during his first seven and a half months as Governor. He also could not reasonably attribute the delay entirely to his predecessor and the Colonial Office because both Young and the Colonial Office had been fully aware of the practical problems referred to by Grantham. The legal problems had in fact been dealt with in the three draft bills, which were completed before Grantham became Governor. The Attorney-General of Hong Kong had gone so far as to disclose to a local newspaper, on the day prior to Grantham's arrival as Governor, the government's timetable of reform. (The draft bills were to be submitted to the Legislative Council in September; the municipal elections were to be held in December; and the council was to be set up on 1 April 1848.[13]) The Attorney-General must, therefore, have considered the legal problems to be largely solved, and, indeed, have obtained permission from the Officer Administering the Government before he publicly committed the government to such a timetable.

The technical problems relating to the establishment of an electoral process had also been tackled adequately by Young and Hazlerigg in an eight-page memorandum. They had even considered such details as the need for Chinese (as well as Western) pens and ink in the polling stations.[14] If the first post-war Urban Council elections, which took place in 1952, are any indication, the technical problems were not as difficult to overcome as Grantham had implied. It required less than three months for the government to reintroduce elections to the Urban Council, including the registration of voters and the final counting of votes.[15] It must be noted, however, that the Urban Council elections, which did not require the delineation of voting wards, involved only 10,000 or so electors, whereas the Young Plan proposed to involve over 100,000 voters.

Although Grantham was correct in his assertion that there was, in March 1948, no suitable candidate for the position of town clerk, the delay between November 1947 and March 1948 cannot justifiably be attributed to this fact. A town clerk would not have been needed until the municipality was about to be established; such legal and technical problems as revision of the draft bills and delineation of the voting wards, for example, were tasks for the legal and other officers (such as Lee, the Assistant Colonial Secretary) of the existing government. In any event, the Colonial Office seems to have had no difficulty in selecting a town clerk for Hong Kong. A Colonial Office minute of November 1948 indicates that several experienced town clerks had been short-listed for the Hong Kong position, but that the Governor had specifically asked the Colonial Office not to interview them.[16] With respect to the implied difficulty in selecting a special officer to oversee the launching of the council, Grantham's appointee responsible for constitutional reform, Lee, had been reassigned in late February 1948 to take charge of the Post Office — less than a month before Grantham complained of the difficulty in finding such an officer.[17]

Grantham was also misleading in his assertion that he had not submitted the draft bills to the Legislative Council because he was waiting for the Secretary of State's comments. The responsible Colonial Office official was justifiably indignant upon reading such a statement in the *Hong Kong Hansard* in early June because Grantham's covering letter to the draft bills had said specifically that 'the drafts were sent for preliminary information, not for specific comment', and that the Colonial Office 'might expect amended drafts' following consideration of the Colonial Office's draft dispatch — which had been sent to Hong Kong in late October 1947 — in the Executive Council. By the end of 1948, a full year later, the Colonial Office still had not received amended versions of the draft bills.[18] When asked in June 1948 to confirm whether he was waiting for comments on the draft bills, Grantham replied that he had hoped to do so, but had then thought that it would be better to defer action until after the draft bills had been formally submitted.[19] For reasons of his own, therefore, Grantham appears to have decided to mark time. For almost eight months, he made no serious attempt to solve the practical problems which he attributed as the cause of the delay. Whether his reasons justified such a policy of procrastination is another matter.[20]

In the spring of 1948, responsibility for Hong Kong was transferred from the Eastern Department to the newly created Hong Kong and Pacific Department under J. B. Sidebotham in the Colonial Office. The progress of reform in Hong Kong was subsequently reviewed. It was recognized that when Creech-Jones had approved Young's scheme in 1947, he had also authorized the reorganization of the Legislative Council and that such reorganization would require the amendment of the Royal Instructions. Preparatory work had already begun, but it had been halted when Mayle recommended in October 1947 that the municipal council be subjected to extensive external control. Grantham had been asked to comment on this matter, but by November 1948 he still had not replied.[21] In the interim, the Indian government had indicated its interest in the Hong Kong reforms and its wish to secure one Indian representative in the new Legislative Council. This request was turned down, however, on the grounds that the Indian population of the colony was not numerous enough to warrant special representation in the Legislative Council.[22]

The Colonial Office also examined at this time the question of whether it should make an unequivocal statement on Hong Kong's future because of the pressure from the Chinese and expatriate British community leaders in Hong Kong. Although it was agreed in the Office that such a statement should be made, Creech-Jones decided in June 1948 to avoid taking any initiative in the matter because of the strong opposition to it by Ernest Bevin, the Foreign Secretary. Creech-Jones was certain that if he were to refer the matter to the Cabinet, Bevin's view that nothing should be said about Hong Kong lest it be provocative to China would be accepted.[23] Political developments in China at the time were not regarded as directly relevant to Hong Kong's reforms.

The situation began to change, however, towards the end of 1948. By then, British officials generally recognized that the military, economic, and political collapse of the Kuomintang government in Nanking was a foregone conclusion. Communist control over China north of the Yangtze river was expected in the not too distant future. Nevertheless, the events were still seen as having little political effect on Hong Kong. The greatest threat to the colony was perceived to be an influx of refugees and — if the Chinese Communists wished to conquer south China — local unrest instigated by Communist-controlled local public utility unions. It was thought that the Communist penetration of south China would take some time, and that even if the Communists were to control south China, Hong Kong would still be allowed to remain under British rule. British diplomats in particular were cautiously optimistic about the future of Hong Kong.[24]

Although the Colonial Office shared the same intelligence sources as the Foreign Office, the Chiefs of Staff, the Commander-in-Chief in South-east Asia, and the Governor of Hong Kong, it took a more serious view of the Communist successes in China. It feared that the Chinese Communists might soon take over the position of the Kuomintang in the colony, and it expected them to be more irredentist and troublesome than the Kuomintang. Officials in the Colonial Office had doubts about the wisdom of proceeding with Young's reform proposals under such circumstances. When Dr S. N. Chau, an unofficial member of the Legislative Council who had also just been appointed to the Executive Council, visited the Office, he said that there was no demand for the reforms in Hong Kong. In view of these two considerations, Sidebotham, who had not been involved with the previous deliberations on the Young Plan or with the 1945 proposals, recommended in November that, subject to the Governor's view, it would be desirable 'to go slow, or indeed to mark time ... with the proposed revision of the constitution'.[25]

When asked to comment on Sidebotham's proposition, Grantham consulted with his Executive Council. The Council thought that the confusion in China would diminish, rather than increase, the danger of Chinese political infiltration in the colony. It also acknowledged that if the government were to renege at this stage, having made public its commitment to reform, political capital would be made by the opponents of the government. However, it was also firmly opposed to the Young Plan. This predicament of the Executive Council was finally resolved when the unofficial members of the Legislative Council (half of whom also sat on the Executive Council) decided to propose an alternative reform (which would involve the complete abandonment of the municipal reform) once the draft bills had been published. When Grantham replied to the Colonial Office, he added that he did not wish to proceed with the selection of a town clerk. He also sought permission to publish four (rather than three) draft bills dealing with the municipal reform. This was the first reference to a fourth draft bill, although Grantham did not indicate what the bill was.[26]

When the Foreign Office was consulted, it replied that, whatever its political complexion, no Chinese government would ever be satisfied with anything less than the restitution of Hong Kong to China. It thought that a liberal reform in Hong Kong would only appease the Chinese government in the short term; it would not change its irredentist attitude in the long term. The Foreign Office therefore thought that the British government need not concern itself about repercussions from the Chinese government if it wished to defer the Hong Kong reforms. However, it also emphasized that if constitutional reforms were to be introduced in Hong Kong, there could be no better time to do so because the confusion in China would make the Hong Kong voters less susceptible to pressure from China. If the Communists had not yet consolidated their control over China, voters in Hong Kong could be expected to vote for neutral or pro-Kuomintang candidates. This situation might change following a total victory of the Communists on the mainland. The Foreign Office itself was indifferent to whichever course of action Hong Kong and the Colonial Office might decide to take.[27]

Grantham's reply was interpreted by Sidebotham to mean that it would be unwise at that time for the government simply to abandon the reforms, but that the unofficial members of the Legislative Council would 'back-peddle' to the extent that the municipal project would in effect be abandoned. Sidebotham was so certain of his interpretation of the situation that he deemed it appropriate to authorize Grantham to publish the four draft bills without checking what the fourth bill was. However, the Legal Adviser in the Colonial Office (Sir Kenneth Roberts-Wray) threatened not to be responsible unless the draft bills were inspected by his department before they were published and this legal scrutiny presumably occupied the Colonial Office for the next three months.[28] Creech-Jones, who had hitherto insisted on a speedy implementation of the reforms, finally agreed to them being reviewed. It was understood that the next move would be made by Grantham and his unofficial advisers.[29]

During this eighteen-month period, there was a conspicuous lack of any real progress in the implementation of the reforms.

The most significant achievement was the drafting of a relatively short dispatch which set out Whitehall's views on external control over the proposed municipality. The failure otherwise to bring reform closer was particularly remarkable in view of the urgency of the matter (which Creech-Jones had stressed repeatedly prior to his approval of the reforms), the non-controversial manner in which the question of external control was being discussed during this period, and the absence of any major practical difficulty in implementing the Young Plan. In fact, the new Governor, the Colonial Office officials responsible, and the unofficial members of the Executive and Legislative Councils all disliked the municipal scheme. Hence, the new Governor avoided taking any real steps towards the implementation of the reforms. Indeed, unlike Young (who had ensured that his Special Adviser for constitutional reform — Hazlerigg — enjoyed direct and frequent access to him through his appointment to the Executive Council), Grantham did not see the need to give his two appointees similar access. He apparently did not wish Lee, his first appointee, to play an important role in the deliberations on the policy for reform. (Lee was not even invited to attend meetings of the Executive Council during this period.) Grantham's attitude is hardly surprising, given his view of the Young Plan and the fact that Lee was one of the least capable of the Hong Kong Cadets, having served twenty-two years as a class two Cadet in Hong Kong without being promoted to class one status, and being one of the few Cadets to be retired at age forty-five.[30] Although Mayle and his assistants did not go quite so far as Grantham, they no longer treated the issue with priority. This change in attitude was possible because Creech-Jones was preoccupied at the time with events in Palestine and could not spare the time to supervise the implementation of a non-controversial policy for Hong Kong — a policy which he had already approved.

In summary, the non-implementation of the Young Plan was due partly to the length of time it had taken both the Treasury and the Colonial Office to reach a decision on the question of external control, and partly to the inadequate communication on the matter between the Governor and the Colonial Office. The crucial factor, however, was simply the Governor's decision (which was strongly supported by his unofficial advisers) to avoid taking the practical steps necessary for the implementation of the reforms. The reasons given for the decision — that the legal and technical issues involved had been underestimated by Grantham's predecessor, and that the Attorney-General had gone on leave for seven months in 1948[31] — were immaterial since the delay during this period cannot be attributed to the government's inability to deal with any of these problems. Whilst the long leave taken by the Attorney-General must undoubtedly have increased work pressure for the Legal Department, it remains difficult to see why the acting Attorney-General could not have handled the legal problems involved in introducing the reforms if any attempt to do so had been initiated during those seven months. Furthermore, there is no evidence that the Governor attempted to deal with any of the legal or technical problems involved in the reforms during this period. Indeed, he omitted for most of this period to appoint an official to be specifically responsible for the reforms. The months between February 1948 (when Grantham reassigned Lee to take charge of the Post Office) and 1 November 1948 (when W. J. Carrie, a class one Cadet, was appointed Special Adviser but, unlike Hazlerigg, without a seat on the Executive Council) coincided with six of the seven months during which the Attorney-General was on leave.[32]

If the Young Plan had been introduced, could it have worked? Given the speculative nature of this question, no conclusive answer is possible. Nevertheless, from the information available, there is no reason to believe that it might not have worked. Even those opposed to the Plan — Grantham and the Unofficials — did not oppose it because they thought it could not have been successful. They opposed the Plan either because they thought that it represented the wrong approach, or simply because they did not wish for any reform. Admittedly, the original timetable made public by the Attorney-General in July 1947 could not have been adhered to because of the length of time required for the Treasury to comment on the question of external control. However, had Grantham followed Creech-Jones' instructions and taken steps to prepare for the introduction of the Young Plan, the reforms presumably could have been successfully implemented in late 1948 or early 1949, at a time when the rival Chinese political parties were too preoccupied with their struggle on the mainland to infiltrate elections in the colony.

An important question remains: was Grantham justified in not implementing the Young Plan? With the benefit of hindsight, the non-implementation of the Plan would appear to have been a mistake since it was workable and provided probably the most appropriate solution in the circumstances. Grantham, however, was strongly opposed to the Young Plan, and he would have failed in his duties had he meekly acquiesced in its implementation: a colonial governor was not supposed to be a 'yes man'. Nevertheless, Grantham was at fault in not putting forward a convincing explanation for his failure to introduce the Plan during the period under review: a colonial governor was also not supposed in effect to put aside carefully considered and duly aproved policies without sufficient grounds.[33] In conclusion, then, whilst Grantham was justified, in principle, in desisting from implementing the Young Plan and in proposing an alternative to it, he was at fault in adopting a policy of procrastination without furnishing satisfactory reasons for such a policy.[34]

Waiting

When Young had returned as Governor in 1946, he was generally welcomed by the articulate people of Hong Kong; Grantham's arrival as Governor was greeted, however, with warmth and enthusiasm. All the major Chinese-language newspapers heartily welcomed Grantham as the first truly post-war governor appointed by the Labour government, and emphasized that the people expected him to lead the colony to a very bright future. On the whole, the local Chinese political commentators had a simplistic view of British colonial policy. They considered the reforms in India and Burma to be primarily the work of the British Labour government. They also thought that this policy of leading British dependencies to self-government was a general policy applicable to all British colonies. They therefore interpreted the announcement of Creech-Jones' approval of the Young Plan, on the day prior to Grantham's arrival, to mean that Grantham had been appointed specifically to carry out such a policy in Hong Kong. In addition, because Grantham had begun his career as a Cadet in Hong Kong and was known to many of the leading citizens of the colony, he was regarded as an old friend who knew Hong Kong and its people well.[35]

With respect to the approved reforms, all the leading Chinese-language newspapers were of the opinion that they could be improved to advantage, but all applauded them as a step in the right direction. The equal division of seats in the proposed municipality drew the most fire, the consensus being that there should be a slight majority of the Chinese on the council. The discrepancy in the qualifications demanded of Chinese and British voters was on the whole considered to be unsatisfactory, and the restrictive qualifications for the franchise were seen as excluding too large a section of the community. The decision to give the proposed council at its inception less authority than Young had originally proposed was also a cause of discontent, as was the Secretary of State's decision to reserve for the time being the questions of financial control and reserve powers. It was pointed out correctly that if the scope of the council were to be less than what Young had proposed and the council were to be subject to extensive external control, it would amount merely to an expanded version of the existing Urban Council. Whilst it was hoped that the opinions of the Chinese community would be respected when the government took the next step towards implementing the reforms, it was also recognized that such an important matter as constitutional reform would entail gradual changes.[36]

The leading English-language newspapers in Hong Kong took a slightly different view of the approved reforms. Creech-Jones' timely approval of the reforms was seen as fortuitous. Although there was some scepticism about the introduction of elections by ballot for the local Chinese, the municipal council was accepted as 'in principle a *fait accompli*'. It was generally agreed that the reforms should be made to succeed. Young was seen as the architect of the reforms; Grantham was hailed as the builder.[37]

The outburst of local newspaper editorials at the time of Grantham's arrival was followed by a lull. The local press considered that its views had been made clear and that the new Governor could be relied upon to take the next step. Such an attitude was appropriate since if a timetable for reform had already been fixed, there would be no need to push the government to proceed with the reforms. Furthermore, Grantham's early public speeches had indicated that the public need not do anything and that the reforms would be implemented without undue delay.

The attitude of the Chinese community remained unchanged at the time of the visit to Hong Kong in March 1948 of Lord Listowel (the newly appointed Minister of State for the Colonies), during which time Grantham spoke publicly of the practical difficulties in introducing the reforms. With the exception of the

Communist *Hwa shiang pao,* all the leading Chinese-language newspapers reiterated their views on the reforms. They requested that the Minister study the issue, consider their suggestions concerning the reforms, and implement Labour's colonial policy, which was generally admired. Significantly, none of the newspapers commented on Grantham's speech about the practical difficulties in introducing the Young Plan.[38] The newspaper which in July 1947 had interviewed the Attorney-General and published the government's timetable for reform continued to anticipate a new move from the government, going so far as to predict that Listowel would make an announcement about the reforms.[39] In fact, Listowel's visit to Hong Kong was for the purpose of familiarization only, and he was not authorized to make any policy statement.[40]

In August 1948, a full year after the Young Plan had been approved, the English-language press began to voice its discontent about the lack of any progress in constitutional reform. The main criticism was that the 'endless delay' in establishing a municipal council had reduced the attractiveness of the venture, and had resulted in the non-implementation of the more important Legislative Council reform. The proposed municipal council had begun to be seen as too restricted in its authority, and the 'enthusiastic reaction' of the people a year ago had turned to 'apathy'.[41]

This assessment by the English-language press of the feelings of the general public about the reforms was not entirely accurate. Whilst the long delay probably caused the scheme to be forgotten by the silent majority — who in any case were uninterested — the Chinese-language press and other interested local organizations did not doubt the government's good faith in this matter. They were merely waiting for the Governor, the 'father and mother official' (and, supposedly, the expert), to make the next move. It was for this reason that Grantham's explanation to the Legislative Council — that the delay in implementing the reforms was due to practical difficulties — went unremarked in the Chinese-language press. (It would be invidious in a traditional Chinese community for the son to criticize the father for failing to do a certain thing because the father found practical difficulties in doing so.) The spontaneous outburst of Chinese-language newspaper editorials on the reforms during Listowel's visit proved that the issue had not been forgotten. In addition, the formation in early 1949 of an embryonic political party, the Hong Kong Reform Club, indicated that the idea of reform was still very much alive in 1948.[42]

Lack of enthusiasm about the reforms was most noticeable amongst the unofficial members of the Executive and Legislative Councils. Between July 1947 and March 1949, they had referred publicly to the subject only once — when Arthur Morse, the senior Executive Councillor, made his welcoming speech on the occasion of Grantham's arrival as Governor on the day following Creech-Jones' announcement of his approval of the reforms. At that time, Morse had assured the Governor of the co-operation of the entire community of Hong Kong, but he had stressed that the community was 'fully alive to the heavy responsibilities entailed by these changes'.[43] Although Morse was genuine in his offer to the Governor of his co-operation on this matter, it did not mean that he and his colleagues were in enthusiastic support of the reforms. If any of the Unofficials had been keen to see the reforms implemented, the extraordinary delay would have been remarked upon either privately in the Executive Council or publicly in the Legislative Council. In fact, the unofficial Councillors avoided the subject in all Council meetings, as did the Executive Councillors, who possibly did not want their views to be recorded had they expressed an opinion. Even Dr S. N. Chau, who had referred to representation when he opposed direct taxation prior to Young's retirement, remained silent on the subject. The available information therefore suggests that the Unofficials supported Grantham's policy of non-implementation of the reforms. Indeed, Dr Chau had actually told Sidebotham privately, when the Colonial Office had begun to doubt the wisdom of introducing the Young Plan, that the reforms were unwanted in Hong Kong.[44]

The primary concern of the Unofficials was to obtain public assurance from the British government that Britain would not desert Hong Kong in the event of a showdown with China. According to Grantham, the British government's refusal to make such a public statement caused the Unofficials to take a generally critical view of the British government.[45] However, in spite of the Unofficials' strong desire for public assurance that Britain had no wish to change the political status of the colony, they were unwilling to press for the implementation of the Young Plan, a step which could only indicate the British determination to stay in Hong Kong.[46]

On the whole, the Unofficials' treatment of the reform issue during Young's governorship and during Grantham's first one and a half years as Governor suggests that they were opportunistic in their attitude. Although they strongly opposed the Young Plan, they did not make their opinion known to Young; they actively opposed the Plan only after the appointment of a Governor who was sympathetic to their views.

Economic Rehabilitation and Reform

The years 1947 and 1948 were ones of rapid commercial expansion in Hong Kong. The total value of trade in current prices reached an unprecedented level of $2,767 million in 1947; in 1948, this figure increased by thirty-two per cent to $3,660 million (more than double the 1946 total of $1,700 million and almost triple the 1940 figure of $1,375 million). Such a phenomenal growth rate represented not merely the restoration of pre-war trade, but an all-round expansion of trade, including that with China. Significantly, however, the proportion of trade with China was considerably less than in the pre-war years. In the five-year period between 1936 and 1940, for example, the annual value of Hong Kong's China trade averaged thirty per cent of the colony's total trade, whereas in 1947 and 1948, China trade accounted for less than twenty-four per cent and twenty per cent respectively. This change in the composition of Hong Kong's trade was a result of the civil war in China and of Hong Kong's efforts to develop new trade and to widen its trading sphere. North America, the British Empire, and East Asia (excluding China) were becoming increasingly important as trading partners.[47]

Hong Kong's post-war industrial development was also considerable. Between 1947 and 1948, 168 new industrial establishments were registered with the Labour Department (a seventeen per cent increase in one year). The number of workers employed in these registered industrial establishments (excluding the unregistered industrial establishments, for which no figures are available) was 51,338 in 1947 and 60,598 in 1948 (an annual growth rate of eighteen per cent). The leading growth industries were textiles, food-processing, printing and publishing, and the manufacture of machinery, metal products, and non-metallic mineral products. During this period, Hong Kong's industries also became more soundly based than previously. Up until late 1947, the most important factor which aided Hong Kong's industries was their ability to offer prompt delivery. By the end of 1948, however, Hong Kong had changed from a seller's market to a buyer's market. As the world supply situation eased, Hong Kong's products faced competition in the world market and there was consequently a fall in the profit margin. However, Hong Kong's industries coped with this new challenge. In 1948, for example, the number of textile establishments fell by four and a half per cent to 387; however, the number of workers employed increased by forty-three per cent to 13,347. The inefficient mills were eliminated and the remaining ones were converted to a larger scale of production. The outstanding industrial growth in the colony during this period, particularly in the textile industry, was fuelled to a significant extent by a huge injection of capital from China. The total value of Chinese industrial capital invested in Hong Kong between September 1945 and May 1948 amounted to HK$100 million.[48]

In addition to the steady inflow of Chinese capital, the rapid economic rehabilitation and growth was also due to general factors such as the political and social stability, the relative efficiency of the colonial government, the restoration of the rule of the law, and the local work ethics — which put the making of money before anything else. Two factors were particularly important in the rehabilitation of the entrepôt trade, which accounted for a significant portion of the colony's total trade. These factors were Hong Kong's status as a free port, and the existence of an open market for American dollars, despite Hong Kong's being part of the sterling area.[49] The fact that most of the countries in the region did not enjoy such benefits also contributed to Hong Kong's spectacular achievement.

The standard of living of the working class in Hong Kong also improved slightly during this period. The cost of living as measured by the retail price index fluctuated, but it showed general improvement. The average retail price index between June 1947 and December 1948 was only 92.57 (compared with 100 when it was first prepared in March 1947). The real improvement was not as impressive, however, as these figures indicate. The average retail price index for 1948 as a whole was 93 (compared with 94 in

1947).[50] The retail price index for food alone remained the same (92) during both years. The actual standard of living of the worker remained very low. The working man in the colony lived at a subsistence level in dingy surroundings. He earned just enough to live on a plain diet, to buy cotton clothes, to rent a cubicle or (more often) 'bed space', to smoke an occasional cigarette, and to send a little money to his family in China. The Labour Department figures indicate that the lot of the workers had not really improved significantly since 1939. Whilst conditions had improved slightly for the artisan, those for unskilled, clerical, and technical workers had not. However, workers were generally more contented in 1948 than in 1947, the number of strikes having declined from twenty-three in that year to five in 1948.[51]

The general economic situation had already begun to improve by the time that Grantham arrived as Governor and the economic rehabilitation of the colony was already well under way. In March 1948, the Financial Secretary of Hong Kong was able not only to project a budgetary surplus, but also to emphasize that government funds would be available for economic construction and development, not just for rehabilitation.[52] By October, Treasury control over the colony had been officially terminated and, by the end of 1948, practically all traces of the war had been effaced.[53]

The rapid economic rehabilitation and development in post-war Hong Kong had little effect, however, on reform during this period. Although the trading community was concerned to some extent about the uncertain future of the colony, it was not compelled after August 1947 to demand either the postponement or the immediate implementation of the Young Plan. The Hong Kong General Chamber of Commerce and the China Association submitted representations on the reform proposals in 1947 not so much because they felt that their economic interests were threatened by the reforms as because they believed that the pre-war Shanghai Municipal Council provided the most appropriate model for local government in a predominantly Chinese city. (Their lobbying efforts would not otherwise have ceased completely after August 1947.) Indirectly, the rapid economic rehabilitation did have some potentially positive effect on the reforms. The colony was becoming prosperous enough to finance the additional expenditure which the establishment of a municipality would have entailed, and the very slight improvement in the local people's general standard of living also created an environment that was relatively more conducive to political development. However, the potential of these changes remained undeveloped because no effort was made to implement the reforms.

The Challenge of the Chinese Political Parties: II

The influence of the Kuomintang in Hong Kong reached its peak towards the end of Young's governorship. Although the Party welcomed any further expansion of its influence over the Chinese community in the colony after Grantham's arrival, it failed to make any significant progress. The only issue on which the Party could hope to gain ready support was that relating to the sovereignty of China,[54] but even in this matter local support did not always materialize in the form which the Kuomintang would have liked. The most notable example of support for the Kuomintang was the so-called Kowloon Walled City Incident — the most important local incident between 1947 and 1949.

The Kowloon Walled City — a small garrison and administrative compound of six and a half acres near the southern tip of the Kowloon peninsula — was originally built by the Chinese in the 1840s following the British acquisition of Hong Kong island. When Britain leased the New Territories in 1898, the Convention of Peking specified that Chinese jurisdiction over the Walled City of Kowloon should continue 'except so far as may be inconsistent with the military requirements for the defence of Hong Kong'.[55] Subsequently, local resistance to the British occupation of the leased territories provided the justification for British forces to expel the Chinese officials and take over the Walled City. This occupation was finalized by an Order-in-Council of 27 December 1899, the exercise of jurisdiction by Chinese officials in the Walled City being deemed by the British to be incompatible with the defence requirements of the colony. However, the Chinese government took a different view of the situation. Although it was in no position to re-establish its authority in the Walled City, it refused to accept the validity of the British act. Mindful of its anomalous legal status, the Hong Kong Police thereafter avoided the Walled City as far as possible, and it gradually developed into a shanty town of vice. When in the 1930s the Hong Kong government declared its intention to evict the local inhabitants in order to redevelop the area, a controversy developed. Despite the British government's doubts about its legal position, it decided to stand by the original Order-in-Council. During the Japanese occupation, the original wall surrounding the City was demolished. By the end of the war, the original Walled City had become a squatter area with unsanitary living conditions and a haven for criminals.[56]

A minor incident concerning the Walled City was initiated in September 1946, during Young's governorship, by Lin Hsia-tze (Lam Hap-tsz, in Cantonese), the Chinese magistrate of Pao An (Po On, in Cantonese) *hsien* (the Chinese county north of the border). After studying the Peking Convention (1898) and other documents, Lin successfully sought the Nanking government's authorization for him to set up a Chinese civil administration in the Walled City. The Chinese Commissioner for Hong Kong, T. W. Kwok, intending to notify the Hong Kong government once Lin had completed his plan, specifically instructed Lin to avoid leaking the plan before he approached Young. However, Lin released the information to the local press before Kwok had been able to take the matter up with Young. Consequently, Young was forced to issue a statement to reaffirm British sovereignty over the Walled City. This statement in turn caused much controversial public discussion. The matter was eventually left to die a natural death following an exchange of notes between the Chinese and British officials which caused considerable embarrassment for Kwok and the Chinese government. On the whole, the Kuomintang did not capitalize on this incident.[57]

The 1947 incident — which bore no direct relation to the earlier incident — was caused by the Hong Kong government's decision to clear the Walled City of its 25,000 inhabitants because the area was wanted for redevelopment, having become a fire and health hazard. That such a decision might be potentially explosive was not recognized, the events of 1946 having been forgotten.[58] On 27 November, the government gave two weeks' notice to the inhabitants of the Walled City to move out and demolish their huts. However, the local inhabitants organized a Residents' Association to oppose this decision. Whilst Chinese patriotism might have been one consideration, the main concerns of the inhabitants were personal: the extension of Hong Kong law into the Walled City would ruin the livelihood of many residents, particularly the common criminals and mainland

Chinese-trained doctors and dentists who could not meet the Hong Kong licensing standards. The residents appealed to the Chinese authority for help. Kwok consequently took the matter up with Grantham and went to Nanking for consultation. In deference to the strong public pressure, the Hong Kong government did not carry out the eviction on the scheduled date. However, the residents of the Walled City were charged with squatting on Crown land.[59]

The attitude of the residents hardened following an address by the Chinese magistrate of Pao An in which Lin publicly declared that the Walled City was Chinese territory. This declaration contrasted sharply with those of Kwok, who attempted initially to persuade the residents to leave the City peacefully. When rebuffed, Kwok dissociated the Chinese government from the likely consequences. The Residents' Association backed Lin, whose declaration was interpreted as implying the Chinese government's support, and Kwok was publicly denounced for adopting a weak-kneed policy. On 12 January 1948, having begun to regard the situation as a challenge to the government, Grantham sent the police into the Walled City to reassert British authority. According to Grantham's account of subsequent events, the police officer in charge 'bungled things' when he opened fire at the squatters. As a result, one resident was killed and dozens were injured; fourteen policemen were also hurt. The Chinese government reacted extremely mildly under the circumstances. Foreign Minister Wang Shih-chieh delivered a 'strong verbal protest' to the British Ambassador, but he later indicated that a satisfactory compromise might be reached. On 16 January, however, events took another ugly turn when one Kuomintang faction, the 'CC (Central Club) Clique', sent its supporters to a massive protest outside the British Consulate-General at Shameen in Canton. The protesters proceeded to burn down the British Consulate and the offices of two British trading companies, wounding in the process several members of the Consulate-General. Foreign Minister Wang, who wished to prevent the Hong Kong question from jeopardizing Sino-British relations, responded with strict orders to play down the Kowloon incident and to prosecute the Shameen protestors. The aftermath of both incidents was left to the diplomats to settle.[60]

Throughout the incident, the Hong Kong Chinese-language press had been critical of the Hong Kong government's heavy-

handed approach. Without exception, all the newspapers expressed regret at the bloodshed in both the Walled City and Shameen.
It is significant, however, that in every case the newspapers
reserved the harshest judgement for Kwok and the Chinese Foreign Ministry rather than for the Hong Kong authorities. More
importantly, the Kuomintang-owned *National Times* was the
most outspoken in this verbal attack. In contrast, the local branch
of the Kuomintang, under Li Ta-chao, remained scrupulously
quiet.[61]

The anomalous response to the incident of the various
Kuomintang organs in Hong Kong is illustrative of the nature of
the Party in the colony at the time. The divergent reactions to the
issue were the result of a factional power struggle within the
Party. At the time of the incident, the 'Political Science Group'
was in majority control of the government, including the Foreign
Ministry. It wished to settle the original dispute through negotiations. The 'CC Clique', however, which was the chief adversary
of the 'Political Science Group', wished to make political capital
out of the incident. The *National Times* (whose editor, Pan
Kung-pi, was close to the 'CC Clique') launched an unrelenting
and bitter attack against the Foreign Ministry and Kwok, in what
was primarily an offensive by the 'CC Clique' against the 'Political Science Group'. When the 'CC Clique' decided to discredit
another of its enemies, T. V. Soong (then Governor of Kwangtung), it engineered the Shameen incident. Li's silence was due to
his close relations with General Wu Te-chen, Secretary-General
of the Kuomintang and a leading member of the 'Political Science
Group'.[62]

Whilst factionalism within the Kuomintang was the chief
source of the Party's weakness in Hong Kong, the deteriorating
fortunes of the Nanking government were the main reason for its
decline in the colony. The primary concern of the Party was to
rally opinion against the Communists, but the local people were
turning more and more away from the Kuomintang, which was
being increasingly crippled in its activities by lack of funds. The
circulation of the *National Times* dwindled to such an extent that
it had to close down in January 1949, and the Kuomintang admitted in a secret document that it was losing ground to its opponents in both cultural and propaganda work. The Party was also
required to abandon its attempt to exercise indirect influence
over the teaching staff and curricula of 250 local schools when

this policy became known to the Hong Kong government. The Kuomintang's influence over labour was also steadily weakening.

By the end of 1948, the governor regarded the Kuomintang as 'no longer a formidable menace to the security and order of the Colony, though they are still capable of a certain amount of mischief, especially if they should choose as a last despairing act to turn to the weapon of assassination, or endeavour to provoke civil disturbance and turn anti-foreign feeling to their advantage'.[63]

The primary task of the Chinese Communists in Hong Kong remained unchanged: to support their struggle for power in China without overtly breaking the laws of the colony. Although the South China Bureau of the Communist Party was located in Hong Kong, the focal points of its activities were Kwangtung and Kwangsi provinces in south China. A secondary function of Hong Kong was to serve as a communications centre for contact with other Communists in South-east Asia. The subversion of Hong Kong was certainly not one of the objectives of the Chinese Communists. Within the colony itself, the Communists continued to confine themselves largely to recruiting new supporters, raising funds, and spreading anti-Kuomintang propaganda. Hong Kong was regarded by the Communists as a major regional, not merely a local, centre.[64]

The Communist attempt in Hong Kong to recruit new young members for the revolution in China met with no real success. The most important recruitment and training centre for young Communist cadres — the Tat Tak Institute — was established in September 1946. From its inception, the Hong Kong authorities suspected that the Institute was a centre for Communist activities and hence denied it registration until December 1947. Although the government had been unable to find evidence to prove its suspicions, the Institute was kept under close surveillance. Intended as a recruitment and training centre for new young members in Hong Kong for service in their home villages inside China, the Institute became primarily a training ground for young cadres from China and South-east Asia. Of the 250 students in April 1948, over half were from Kwangtung province and one-third were from South-east Asia; only ten per cent were locally enrolled. Notwithstanding its success in providing training facilities, the Institute failed as a recruiting agency (as did other similar Communist schools in the colony). The cause of this failure, according to one Communist study at that time, was that the young people of Hong Kong suffered from 'Hong Kong head': an ideological syndrome which included arrogance, a selfish and city-oriented view, as well as a tendency to forget their Chinese identity and to despise their own culture. The Communists suffered a further setback in September 1948 when a routine police street search led to the capture of Communist documents which incriminated, amongst others, two teachers of the Institute. The incriminated teachers were subsequently deported and the Institute was closed down in early 1949.[65]

Unlike the Chinese Democratic League and the Kuomintang Revolutionary Committee (the so-called third force in Chinese politics then in exile in Hong Kong), the Communists invariably avoided embarrassing or attacking the British through their propaganda. Despite its anti-imperialist tone, the Chinese Communists in Hong Kong made no hostile comments about 'British imperialism', with (as far as the Colonial Office was concerned) one significant exception: their support of the squatters in the Kowloon Walled City Incident.[66] The Communists' support of the squatters did not necessarily mean, however, that they were hostile to the government. The eviction was carried out in a very heavy-handed manner and the squatters' cause was strongly supported by the educated Chinese, the prime target of Communist propaganda. If the Communists had remained silent, they would have completely discredited themselves. In any event, and under the circumstances, the Communist newspaper editorials were very mild towards the British. Prior to the forceful and bloody eviction, the *Hwa shiang pao* had restricted itself to criticizing the Kuomintang government; after the bloodshed, it merely regretted the incident, saying that the Hong Kong government had been wrong in resorting to force instead of persuasion. Therefore, whilst the Communists supported the cause of the squatters, they were not hostile to the government. Such remarkable restraint indicates that they had not really departed from their general policy.[67]

As the tempo of the Communist success in the Chinese civil war increased towards the end of 1948, the question of whether the Communists would continue their appeasement policy towards the British in Hong Kong became relevant. In November 1948, Chiao Mu (or Chiao Kuan-hua), Head of the Communist Hsin Hua News Agency in Hong Kong, took the initiative in reassuring the British that the Communist policy and approach would not change. He told H. C. Bough, the Reuter correspondent in Hong Kong, in an 'off-the-record' interview, that his colleagues numbered 'only a few dozen' and that they had no intention of causing any trouble. Chiao made it clear that the Chinese Communists could have perfectly normal relations with the British Labour government. The Chinese Communist Party had no intention of taking Hong Kong by force; it was a diplomatic issue and a very minor one at that. Chiao in fact paid tribute to the British policy of neutrality in the Chinese civil war and he appreciated the hospitality extended to his Party in Hong Kong. He welcomed the British policy of not allowing Hong Kong to be used as a base for operations against the government of China. He also stressed that if Kuomintang leaders were to become refugees in Hong Kong in the future, his Party would understand the Hong Kong government's policy with regard to them.[68] In spite of the Communists' good behaviour and Chiao's reassurances, however, the British government thought that, on the whole, it would be prudent to guard against a possible Communist challenge.[69]

With respect to labour and the trade unions, Brian Hawkins, Head of Hong Kong's Labour Department (and a Cadet), considered that they should not be entrusted with matters of responsibility, particularly in regard to elections. The pre-war trend towards the development of healthy, independent labour unions had been reversed by the Japanese; after the war, the local unions were merely the tools of either the Kuomintang or the Communists. Under such circumstances, if Hong Kong's unions were to be treated in the same way as British unions, it would destroy 'what slender chances there [were] of healthy development' of trade unionism in the colony.[70]

The Labour Department under Hawkins did not in fact pursue an active policy to combat political infiltration of the labour unions by the two Chinese political parties. It adopted instead a policy which is best described as positive non-intervention. In essence, this meant that advice and assistance were given to those workers who wished to develop independent unions, on condition that they took the initiative in asking for such advice and assistance; workers were left to work out their own form of unions. Whilst in principle this was no doubt the best approach to developing a healthy and independent trade union movement, it erred on the side of being too passive in the circumstances and particularly in the light of Hawkins' own observations about the leadership problem in the labour movement. Hawkins observed that it was 'extremely difficult to find among this class of worker men who possess[ed] the necessary intelligence, education [which was defined, in a footnote, as knowledge of English] and organising ability to lead the union[s], but when found they [were] usually politically minded and with definite political ties'. In view of this assessment and the fact that not all of the anti-Kuomintang unions were necessarily pro-communist — many were 'genuine workers' organisations organized on an industrial ... basis' — the Labour Department might have assisted in the development of an independent labour movement if it had applied a more vigorous policy which assisted union officials to acquire a better education and encouraged them to concentrate on improving the lot of the workers instead of becoming involved in Chinese politics.[71] Hawkins seems not to have realized that if the two Chinese political parties were infiltrating the local labour movement, a passive policy of positive non-intervention would not have been able to stop such infiltration.

Although from 1947 onwards the local unions tended to amalgamate, the Labour Department chose to remain primarily a spectator. Hawkins favoured non-intervention, not because he thought it wrong to impose British trade unionism on the Chinese in Hong Kong, but because he thought a trade union congress on the British model was not possible in the colony. He predicted that the labour movement in Hong Kong would eventually produce two loose federations: one pro-Communist and the other pro-Kuomintang.[72] Although Hawkins was proved correct, his policy was an important factor which contributed to the setting up of two rival federations: the left-wing Federation of Trade Unions and the right-wing Hong Kong and Kowloon Trade Union Council.

Whilst it is impossible to predict what the result would have been if the Labour Department had actively supported an independent labour movement, Hawkins and his Department certainly made no attempt to assist its development. During this period, the local labour leaders who wished to develop an independent labour movement looked to the government for direction. The 'pro-Kuomintang' Chinese Engineers' Institute had taken the first step towards seeking an independent path when it rejected the

Kuomintang's offer of financial assistance during its strike of August-September.[73] The first Secretary-General of the right-wing Trade Union Council also welcomed advice from Ken Baker, the Government Trade Union Adviser, particularly in matters relating to the relations between the local labour movement and the 'international labour movement'.[74] Ken Baker, who was sent by Creech-Jones to help develop a healthy trade union movement in Hong Kong, was unable to give much assistance, however, because of constant interference from his superior officer — a Cadet (hence, either Hawkins or his deputy, Q. A. A. MacFadyen) — who, according to Baker, did not really understand labour problems.[75] At the same time, neither the Kuomintang nor the Communists attempted to cut each other's throat in the labour field during this period. The unions which supported either party were not publicly critical of each other. The dividing line between these two groups was 'not always in evidence, as a number of the unions from both sides co-operate[d] with each other on matters affecting them collectively'.[76] Furthermore, only two of the 141 unions publicly admitted allegiance to either of the Chinese parties. The Kuomintang-published *National Times* went so far as to applaud the inauguration of the leftist Federation of Trade Unions in September 1948 as 'a big step forward in Hong Kong's labour movement'.[77] If there had been any chance of developing an independent labour movement in post-war Hong Kong, this would have been the time. However, the Labour Department made no attempt to seize this opportunity.

Between mid-1947 and the end of 1948, neither the Kuomintang nor the Chinese Communists posed a direct challenge to the government of Hong Kong; they remained, as they had in the past, potential threats. Even the Kowloon Walled City Incident, which might have been considered a challenge on the part of the Kuomintang, was caused by the ineptitude of the Hong Kong government itself. On the whole, the local political situation provided no cause for apprehension, and Governor Grantham did not find it alarming. In so far as the threat to political reform was concerned, the situation became more, rather than less, conducive to reform as the Chinese political parties became increasingly engulfed in their life-and-death struggle within China. There is nothing to suggest that if Young's scheme had been implemented, it could not have succeeded in the ways Young had envisaged. Indeed, both the Kuomintang and the Communists disregarded for the most part the proposed Hong Kong reforms after the summer of 1947; their hands were full with other more important concerns. Grantham and the Foreign Office were aware of this situation when Grantham, on the advice of his Executive Council, proposed an alternative to the Young Plan at the end of 1948. The Colonial Office agreed to consider the alternative plan mainly because of the prospect of having to face the Communists as the major threat in the colony.

5 Metamorphosis

In early 1949, the Governor and the Unofficials co-operated in their moves to put forward an alternative to the Young Plan. However, they contrived to make it appear that the Unofficials had taken the initiative entirely on their own, presumably because as representatives of the people (albeit only appointed and not accountable to the general public), the Unofficials did not wish to be seen as 'yes men' or as being in collaboration with the Governor in this matter.

The Governor made a discreet start when he announced in mid-March, in his annual address to the Legislative Council, that the draft bills required for the municipal council reform were soon to be published for public information and comment. D. F. Landale, the senior unofficial member of the Legislative Council, responded during the following meeting of the Council. He doubted whether the people of Hong Kong wished for the introduction of a municipal council. In the next meeting, Grantham publicly assured Landale and the people of Hong Kong that the government did not intend to steamroller the existing proposals through the Legislative Council, and he openly invited the Unofficials to made counter-proposals. The first step in the burying of the Young Plan had thus been taken.[1]

Four weeks after Grantham's call for counter-proposals, Landale gave notice, on behalf of all the Unofficials in the Legislative Council, of a motion to put forward an alternative to the Young Plan. He said that for the last year or more, the Unofficials had been sceptical of the Young Plan and had 'made enquiries as to whether His Majesty's Government were prepared to consider alternative proposals'. He added that the counter-proposals were being made at that time because 'the first indication' that Whitehall would consider an alternative reform had been made public by J. J. Paskin, the Assistant Under-Secretary supervising Hong Kong affairs, when he visited the colony in January 1949.[2] This rationalization contradicts the available evidence, however. There is no record of any Unofficial taking the initiative before 1949 to enquire about the view of the British government; Paskin's January statement was the confirmation, not the 'first indication', of Whitehall's attitude.[3]

On behalf of all the Unofficials, Landale proposed that Young's municipal council scheme be abandoned and that his recommendations regarding the Legislative Council be replaced by a different set of suggestions. Landale and his colleagues asked for a Legislative Council with twenty instead of seventeen members, consisting of nine official members (including the Governor as President) and eleven unofficial members (to be partly elected by British subjects and partly nominated by the Governor), resulting in an Unofficial majority of two. Before the Legislative Council debated the motion on 22 June, Landale left the colony; his role was taken over by Sir Man-Kam Lo. At the beginning of the debate, Lo and his colleagues amended Landale's motion in one important area. They proposed that the Unofficial majority in the suggested Legislative Council be increased from two to five by reducing the number of official members from nine to six, constituting a Council of seventeen (instead of twenty) members.[4]

The Legislative Council 'debate' was not, strictly speaking, a debate since all the Unofficials supported the motion and had some kind of an understanding about what they planned to say; at the same time, the official members had been instructed not to participate. The Unofficials' strategy was that Lo, as leader of the Unofficials on this occasion, would put forward a general defence of the scheme, while the other Unofficials would elaborate on specific issues.

Lo began the 'debate' by disarming potential critics outside the Council. He stressed that opinions appearing in the press should not be relied upon too closely because they indicated the failure of the articulate public bodies to reach an agreement on the crucial question of how to reform the government. Moreover, since all of those bodies combined could not claim to represent more than 10,000 of the local inhabitants, they did not represent any substantial section of the community. He then argued that nominated Unofficials such as himself, who had consulted various public bodies but who were not tied to any one of them, represented the best interests of the colony as a whole. Oddly enough, Lo also said that, when taken together, the opinions articulated reflected that 'the majority' supported the alternative that he and his colleagues had put forward.[5] With the acknowledgement that it is not possible to prove whether or not the opinions articulated represented the view of the silent majority, it must be emphasized that this last assertion by Lo was at variance with the opinions expressed in the newspapers surveyed. The overwhelming majority of the articulate Chinese organizations in fact favoured a different plan which provided for a combination of both Young's municipal scheme and the Unofficials' Legislative Council proposals.[6]

In rejecting Young's municipal scheme, Lo said that he was not rejecting the idea of a municipal council; he was opposed to the immediate establishment of such a council simply because he thought that it would lead to considerable duplication of offices between the colonial and municipal governments and that it would cause an increase in government expenditure. He preferred the gradual expansion of the existing Urban Council — which, in Grantham's words, were merely 'the agent of the central government for certain urban services' — into a municipal council.[7] According to Lo, the advantage of his approach as compared to Young's was that a committee of the proposed new Legislative Council could plan to avoid the duplication of offices.[8] However, Lo was mistaken in his assertion that Young's municipal scheme would inevitably lead to the duplication of offices because, as Lo himself recognized, at its inception Young's proposed municipality would only take over the functions of the existing Urban Council and would have clearly defined powers over the Fire Brigade, the parks, gardens, and recreational grounds, the licensing of vehicles, and the licensing and control of amusements; there would be no duplication of offices at that stage. As to the prevention of such duplication at a later stage, Lo's idea was not fundamentally different from Young's recommendation, which was for a government commission to do the same job that Lo had assigned to a committee of his proposed new Legislative Council. The only important difference between Young's and Lo's proposals with respect to municipal matters was that whilst Young wished to establish a municipal council immediately, Lo preferred to defer the matter for the consideration of his proposed new Legislative Council, which could decide not to set up a municipal council at all. In view of this difference, and the fact (as discussed in Chapter 4) that the Unofficials had reached an understanding in December 1948 with Grantham and the Colonial Office to put forward an alternative which would in effect cause Young's municipal scheme to be abandoned, Lo must have referred to the possible future development of the Urban Council in order to justify the abandonment of Young's municipal scheme to those members of the general public who were interested in the matter. This task may have been facilitated to an extent by the confusion at that time in the standard Chinese translations between the Urban Council and a municipal council, which were very similar. The Urban Council was translated as the 'urban sanitary council' *(Shih-chêng-wei-shêng-chu)*, whereas 'municipal council' was translated as 'urban council' *(Shih-chêng-chu)*.

With regard to the electoral arrangements, Lo and his colleagues proposed that the vote be restricted to British subjects and that communal representation be introduced. Of the eleven unofficial members, six should be Chinese and five should be non-Chinese. Of the Chinese members, four should be elected and two should be nominated by the Governor; of the non-Chinese members (one of whom should be Portuguese), two should be elected and three should be nominated. The Governor should have an original and a casting vote as well as the usual reserve powers. These arrangements represented a move towards the usual approach to constitutional development in British colonies — an approach which had been implicitly rejected by the British government in 1946 as inappropriate for Hong Kong. Hall and Young's announcement of 1 May 1946 had specifically said that the object of the Hong Kong reforms was to give the local inhabitants, not just the British subjects, a share in the government. Lo defended his proposal on the grounds of expedience: if the vote were to be given to non-British subjects it would involve 'interminable argument and delay'; communal representation was inevitable because without it 'a non-Chinese candidate receiving 100% of the votes of the non-Chinese electors might be defeated by another non-Chinese candidate receiving the support of Chinese electors only'. Lo also pointed out that the allocation of seats between the Chinese and the non-Chinese communities was based on the 1931 census.[9] However, this allocation was deceptive. The number of seats given to each community was not proportional to its share in the population. According to the 1931 census, there were 370,000 Chinese who were locally born (and therefore British subjects), but only 61,604 of those had registered as British subjects. Lo proposed that the franchise be limited to those Chinese who had registered as British subjects, and he suggested that they be given six seats in the new legislature. This division of seats contrasted with the four seats allocated to the 6,636 expatriate European British, the single seat allocated to the 1,089 Portuguese British, and the failure to allocate any seat to the 3,331 Indian British. If the 1931 census figures had in fact been used as the criterion for the division of seats, there was no logic behind the arithmetic. Lo and his colleagues also failed to explain their decision that four Chinese and two non-Chinese members should be elected. In fact, the division of seats was based on political and practical considerations. Grantham later explained confidentially to the Colonial Office that the number of seats was so divided that the potentially 'unreliable' members (mainly the Chinese and the elected members) would be unable to defeat the government on matters relating to the political status of the colony.[10]

Lo also stressed that the greatest merit of the Young Plan — the fact that it was to serve as a 'prelude to more substantial constitutional reforms' — would be negated if their recommendations were accepted.[11] This assertion that the Unofficials' scheme represented a more substantial reform than Young's is misleading. When the alternative proposal had first been made (as Landale's motion) it had provided for an Unofficial majority of two, only marginally more than Young's supplementary Legislative Council proposal which provided for an Unofficial majority of one. Lo proposed that the Unofficial majority be increased to five, but this suggestion was presumably made to satisfy the articulate public who had demanded a more liberal reform. For Lo to argue that his alternative was the 'more substantial constitutional reform' on the grounds that it would introduce an element of direct election into the Legislative Council is hardly less misleading since he had made British nationality a condition for its introduction. Such a move would, if implemented, have reduced the number of potential voters from several hundred thousand to, in Lo's estimate, approximately 74,000. (According to the presumably more accurate estimate of the Secretary of State, the number would have been only 16,000.) Such an eventuality had been one of the most important reasons why the British government rejected in 1946 the usual approach of restriction of the vote to British subjects.

At the completion of his speech, Lo appealed to those who

were demanding that 'all members of the Legislative Council must be elected'[12] to adopt a realistic approach and a compromising attitude. He emphasized that the Unofficials were in unanimous agreement that, whilst their alternative may not satisfy completely the aspirations of all the inhabitants of the colony, it represented 'a fair and acceptable compromise'.[13]

The seconder of the motion put forward by Lo, M. M. Watson, stressed the fact that the Unofficials had consulted with the leaders of the community before drawing up their proposals. He said that since the silent majority had no strong preference either for the Young Plan or for the Unofficials' alternative proposal, which was supported unanimously by the Unofficials as 'the logical and safest course', he therefore recommended that the alternative plan be adopted in the colony. He further argued that the introduction of the Young Plan would create the danger of the proposed municipal council, which would consist of both directly and indirectly elected members, challenging the Legislative Council, which would have only appointed and indirectly elected members.[14] Watson was correct in his assertion that the alternative plan could prevent any such confrontation between the proposed municipal council and the Legislative Council for the simple reason that there would not be a municipal council. Watson should also have elucidated why Legislative Councillors should not be subject to questioning outside the Legislative Council. His decision to evade such an important point suggests that his main concern was to protect the vested interests of the unofficial Legislative Councillors.

The Portuguese representative in the Legislative Council, Leo D'Almada E Castro, emphasized in his address the need to secure the representation of the Portuguese on the new council. He pointed out that the expatriate British were merely birds of passage and that the Chinese were too closely tied to interests in nearby China; only the Portuguese, who were the really long-term residents of the colony, had its best interests at heart. D'Almada E Castro also said that the reform movement in Hong Kong was created by the British government and that its main advocates were newcomers to the colony; the 'large majority of residents of long standing in this Colony' (presumably, the Portuguese) did not wish for any reform at all.[15] Although he did not specify who the 'newcomers' were, they would have had to be 'new' in relation to the Portuguese community's long history of settlement in the colony; he could not reasonably have been referring to the Chinese refugees who had arrived during the post-war decade but who were yet to arrive in any significant numbers.[16] He then supported the motion without criticizing the Young Plan or the draft bills.

The representative in the Legislative Council of the Hong Kong General Chamber of Commerce, M. C. Blaker, backed the motion unreservedly. The Chamber by this time had finally put aside its original idea of transplanting the pre-war Shanghai Municipal Council into post-war Hong Kong. Blaker had only one argument: the establishment of a municipal council as outlined in the draft bills would be a 'Herculean task'; the reform of the Legislative Council as outlined in the alternative proposal would be 'far simpler' and 'would denote a definite measure of progress towards giving the citizens of Hong Kong' greater local self-government. He failed to elaborate further, however.[17]

Lo, Watson, and D'Almada E Castro all shared Blaker's view that it would be simpler and faster to introduce their alternative than to implement the municipal scheme. This view contradicted Grantham's observation, expressed earlier in the Council, that the municipal reform could be quickly effected by introducing the draft bills which had already been tabled, whereas reform of the Legislative Council would require amendment of the Royal Instructions, which was beyond the control of the Hong Kong government.[18] Indeed, the introduction of the Legislative Council reform would mean starting from scratch: the proposals would have to be submitted to the Secretary of State and, if approved, the necessary legislation would have to be drafted — two stages which the Young Plan had already completed. Neither Lo, Watson, nor D'Almada E Castro elucidated their assertion; nor did they explain why they thought that the introduction of direct elections into the Legislative Council would be easier than introducing direct elections into the proposed municipal council.

The senior Chinese unofficial Councillor, T. N. Chau, pointed out that 'certain sections of the Chinese community desire[d] not only that this Council be enlarged, but that Hong Kong should also have its long-awaited Municipal Council'. In a bid to placate these sections of the Chinese community, Chau asked the government to appoint a commission to study the possible development of the Urban Council after reform of the Legislative Council. He did not specify the size of these 'certain sections' of

the Chinese community, but he must have thought them to be of
a significant size and/or importance to require consideration.
Nevertheless, Chau supported the motion on the grounds that he
felt that 'the majority of the public' favoured the alternative over
the Young Plan. He did not explain, however, how he was able
to ascertain the views of the silent majority.[19] The content and
tone of Chau's very short speech would seem to indicate that he
did not wish to become involved in the 'debate' at all.

The remaining Chinese Unofficial, S. N. Chau — a cousin of
T. N. Chau — also made representations on behalf of those
certain sections of the Chinese community referred to by T. N.
Chau. He supported Chau's suggestion that a commission be
appointed to examine the development of the Urban Council. He
also emphasized that restriction of the electorate to British
subjects, as was being proposed, was not as equitable as it could
be in the circumstances. His insistence that the non-British
Chinese taxpayers be included in the electorate was, by
implication, in opposition to the motion, yet he ended his speech
by stressing that he had 'much pleasure in supporting the
motion'.[20]

At the end of the meeting and according to plan, all of the
Unofficials, including S. N. Chau, voted unanimously for the
motion, which was therefore carried. However, the Governor,
who like some of the Unofficials resented S. N. Chau's unexpected opposition in his speech to the electoral arrangements
which had been agreed earlier, ridiculed Chau's self-contradiction
when he concluded the meeting.[21] Immediately after the meeting,
Chau called on the Governor in order to explain that 'while he
was personally in favour of the motion he had been asked to
make known the views of the non-British Chinese community'.[22]
His self-contradiction had probably been difficult to avoid: as a
nominated Chinese Unofficial, he had to make representations
on behalf of the Chinese community, while at the same time he
had to vote for what he himself believed to be best for the
colony. It is not clear, however, just what Chau thought was best
for the colony. Did he really believe that the alternative was most
appropriate for Hong Kong in the circumstances? After all, it had
been he who (as discussed in Chapter 4) had told Sidebotham
towards the end of 1948, just before the scheme to abandon the
Young Plan was set in motion, that reform was unwanted in
Hong Kong. Did he in fact believe that the colony would be

better off without any reform? Or had he changed his mind between the end of 1948 and June 1949? The circumstantial evidence and the recollections of people who knew Chau indicate that he very probably simply did not want any change. It is very likely that he openly opposed the restriction of the franchise to British subjects at the Legislative Council 'debate' merely in order to gain the goodwill of the articulate Chinese pressure groups, in case he needed it in the future.

Their speeches in the Legislative Council 'debate' indicate that the Unofficials clearly recognized the strong demand in 1949 amongst the various articulate local organizations for a liberal reform. Because of this recognition, and because of their determination not to be seen as conservatives opposing reform, the Unofficials attempted to supersede the Young Plan by putting forward an apparently more fundamental and progressive reform. The fact that they had refrained from commenting both before and after the 'debate' suggests, however, that the Unofficials could not have been enthusiastic about reform. If any Unofficial had been keen to make the alternative proposal a reality, he would have advocated it to the 142 public bodies which were clamouring for reform in Hong Kong during the spring and summer of 1949, or questioned in the Legislative Council sometime after the 'debate' the lack of progress with regard to the reforms. With the exception of Lo, who as Legal Adviser to the Hong Kong Chinese General Chamber of Commerce had virtually advised the Chamber to accept the Unofficials' alternative, no Unofficial took either of these steps. The alternative proposal put forward by the Unofficials was merely the price to be paid for abandoning the Young Plan.

In late August 1949, the Governor reported to the Secretary of State and recommended that the Unofficials' alternative be accepted, subject only to three additional restrictions on the franchise. In addition to the requirement that all voters be British subjects, voters should also have reached the minimum age of twenty-five; they should have resided in the colony for at least one year since attaining the age of twenty-one; and they should be literate in the English or Chinese language. Grantham stressed that the restriction of the vote to British subjects would encourage the locally born inhabitants of the colony to throw in their lot with the government and register as British subjects. He also advised the Secretary of State to make his decision a final one, after consultation with him if necessary, and 'in no way subject to reconsideration in the Colony': an indication that despite Grantham's public rhetoric, he did not wish to encourage public discussion of the reforms. He emphasized that although it would normally be right and proper to consult extensively the colonial people on such an issue as constitutional reform, Hong Kong was a special case: the vast majority of Hong Kong's population were not British subjects and, as a matter of 'fundamental principle', matters of state in a British territory should be dealt with only by British subjects. More importantly, the Governor thought that the local people would accept whatever final decision the Secretary of State might make.[23]

The Governor also recommended that a commission be appointed to study the development of the Urban Council after the Legislative Council reform. He said that he was opposed to the establishment of a municipal council not in principle, but on practical grounds: a municipal council should not be set up before all of its duties had been very clearly defined. He pointed out that with regard to public works and health, the division of power between the central and municipal governments would cause considerable problems. Since Young's proposed municipality would be responsible only for the urban areas, but not for the rural areas, the central government would need to retain its Public Works and Health Departments, hence causing duplication of offices. Furthermore, since the Police Force was not to be subject to the control of the proposed municipality, it would be difficult to determine what percentage of rates should be assigned to the municipality.[24] Whilst these problems were real and difficult, Young had also recommended that a commission be set up to tackle them before his proposed municipality took over duties in those areas, a suggestion not dissimilar to Grantham's. Nevertheless, Grantham's approach did have one potential advantage over Young's scheme. If the commission were to conclude that these technical problems could not be solved satisfactorily, the government could then decide not to set up a municipal council; this decision could not of course be made if Young's scheme were to be implemented. Since Grantham would be held responsible if Young's proposed municipality were to fail, he was justified in asking for caution — even though the possibility that Young's municipal scheme would fail merely because these technical problems could not be solved satisfactorily was a remote one.

The advantage of the Unofficials' alternative, according to Grantham, was that it avoided the political danger inherent in Young's municipal proposal: the danger that pro-Communist elements would be elected into the municipal council. In this matter, Grantham's concern coincided with that of the Unofficials, although they were not identical. Grantham rejected the idea of a municipal council because it could be infiltrated by the Communists. The Unofficials rejected it because it could be dominated by strong-minded people unfriendly to them — an independently minded (that is, neither pro-Communist nor pro-Kuomintang) municipal council could challenge the Legislative Council as much as a Communist-dominated one could. With respect to the danger that Chinese Communists might infiltrate the proposed new Legislative Council (which was to have directly elected members), Grantham proposed two safeguards: the restriction of the vote to British subjects and the allocation of seats. The rationale behind the former proposal is self-evident. The latter proposal was intended to counterbalance the potentially dangerous element — the Chinese elected members — by the more reliable elements: the four elected Chinese members would be balanced by two nominated Chinese, two elected European, and three nominated European members. In the event that the two nominated Chinese members were forced to support the four elected members, they would still be outnumbered by the five European and five official members. If the worst came to the worst and the two elected European members were to vote with the six Chinese members, the government could still rely upon the three nominated European members and the five official members, as well as the Governor (who had an original as well as a casting vote). There was therefore no possibility of the Council being forced by the Chinese Communists or by other outside political forces to carry a motion against the Hong Kong government.[25]

Grantham also commented on Creech-Jones' personal view on reform in Hong Kong, which he had expressed in early 1947 while Young was still Governor. He thought that Creech-Jones' quasi-ministerial proposal was inappropriate for Hong Kong because 'some of the elected members of [the] Legislative

Council, whilst nominally giving allegiance to the Crown, would in their hearts be giving allegiance to a state [that is, China] which wishes to end British rule in the Colony'. Grantham thought that it would be difficult to avoid appointing some of those people, who could claim to be 'true representatives of the people' to the Executive Council if a quasi-ministerial system were to be introduced. He defended communal representation against Creech-Jones' attack — given the racial composition of the population in Hong Kong, the failure to have communal representation would mean that British subjects of the Chinese race could capture all the seats in the Council.[26] He also stressed that it would be essential for the Governor to continue to preside over the Legislative Council. Grantham, aware that Creech-Jones would probably disagree with his recommendations, perhaps wished to forestall Creech-Jones from raising his old arguments.

Grantham also at this time assessed the opinions expressed by the local newspapers and various local organizations on constitutional reform. He reported that the English press in particular criticized Young's municipal scheme for 'attempting too much and offering too little'. However, they were also dissatisfied with the alternative proposed by the Unofficials. The newly formed Hong Kong Reform Club, which was constituted by professionals of many nationalities, wished to widen the franchise proposed by the Unofficials to include all 'Hong Kong citizens'. Grantham considered the Club to be misguided. He thought that all of those long-term residents who did not register as, or naturalize in order to become, British subjects would never have the colony's best interests at heart. His advocacy of communal representation suggests, however, that he distrusted even those local Chinese who had thrown in their lot with the British by claiming British nationality. The Chinese press in general endorsed the views of the 142 organizations that had petitioned the Governor to combine Young's municipal proposal with the Unofficials' alternative (both suitably amended). This petition indicated that while they thought the Young Plan offered too little, these organizations did not consider that it attempted too much.[27] Grantham dismissed their representation for two reasons: no evidence had been adduced by the Chairmen of these bodies to show that they were 'authorized by their respective associations to sign the petition', and some who had signed were 'recent arrivals'. He further argued that these Chinese organizations were unduly influenced by the history of the pre-war Shanghai Municipal Council, and that they failed to appreciate the difference in status between Shanghai and Hong Kong. It is to be regretted that Grantham did not elaborate on this assertion because such information as is available suggests that it was the expatriate British, rather than the local Chinese, who had been strongly influenced by the pre-war Shanghai experience. Indeed, the overwhelming majority of the local Chinese leaders in Hong Kong at that time were Cantonese, and it is doubtful whether they would have had much knowledge of the pre-war Shanghai Municipal Council. Grantham also mentioned that there was a strong outcry amongst the articulate sections of the community for taxation with representation, which had been promised by Young when he introduced income tax in 1947. He even conceded that 'political consciousness has been awakened in Hong Kong'. He remained firm in his belief, however, that the articulate opinions were misguided and that they should not be too heavily relied upon.[28] Having had over twelve years' experience in the colony before the war, Grantham was confident that he knew what was best for Hong Kong. He was in full support of the Unofficials' alternative because it had been drawn up under his own guidance.

While constitutional reform was being reviewed and debated in Hong Kong during the first eight months of 1949, the British government was becoming increasingly concerned over the security of the colony and British interests inside China as a result of events on the mainland. It had been apparent from late 1948 onwards that the Communists were winning the Chinese civil war and the British government had begun to assess the implications which their victory would have for the security of Hong Kong. The Chiefs of Staff Committee in early 1949 assessed the most serious threat to the colony, which might occur before the third quarter of the year, to be a vast influx of refugees; an additional threat was thought to be Communist-inspired strikes. The Committee was not of the opinion that the Chinese Communists would invade the colony, but it considered that the local security forces would need to be reinforced by a brigade group if they were to be capable of coping with both an influx of refugees and a Communist guerrilla attack from across the border.[29] The Colonial Office took a more serious view of events than did the Chiefs of Staff Committee. It thought that the Chinese Communists could inspire strikes in the colony 'at any time'. It also believed that a refugee influx might occur in the next six months and that the security forces 'available and being formed... [were] wholly inadequate to meet a serious threat'. The Colonial Office wished to strengthen the local garrison as much as possible, but it failed to obtain Cabinet support on this matter when the latter examined in March 1949 the security problems of the colony.[30] Of all the British officials responsible for Hong Kong (including those in Hong Kong, China, and South-east Asia), those in the Colonial Office were the most anxious about the possible Chinese Communist threat to the colony.

The British government also reviewed at this time its China policy, the main thrust of which in the immediate post-war years had been to protect British investments and other interests inside China. The value of British properties inside China had been estimated in 1941 at approximately 300 million pounds, almost double that of British investments in Hong Kong (including the investments of the locally registered British *hongs*). Subsequent war damages had no doubt reduced the figure, but British interests in China in 1949 remained substantial, and 'most of the property could not be got out'. The Foreign Office considered that the best hope for British interests in China was to maintain them in the country for as long as the Communists allowed them to do so. It also predicted that the Communist or Communist-dominated government, which was about to be formed in China, would still have to rely for some time on the British-run public utilities, insurance, banking, commercial, and shipping agencies, and industrial enterprises. Consequently, the British Cabinet decided to adopt a policy of 'keeping a foot in the door'. This policy meant that provided there was no actual danger to life, British personnel 'should endeavour to stay where [they were], to have *de facto* relations with the Chinese Communists in so far as these [were] unavoidable, and to investigate the possibilities of continued trade in China'. It was hoped that Britain could eventually 'insist as a *quid pro quo* that the Communists should respect [the British] trading position and [British] properties in China' by persuading other Western countries to act in concert with Britain in the supply of essential materials to a Communist-dominated China.[31] In general, then, the British government was more concerned in early 1949 with British interests inside China than with the security problems of Hong Kong.

The Chinese Communist threat to Hong Kong was dramatized by an armed clash between the Communist Army and the Royal Navy. On 20 April 1949, a British frigate, HMS *Amethyst*, which was on its way from Shanghai to Nanking along the Yangtze river, was fired upon by Chinese Communist troops who were preparing to launch an attack against the Kuomintang forces across the river. The frigate was holed and grounded, and her captain, Lieutenant-Commander B. M. Skinner, later died of his injuries. Subsequent attempts by the British Far Eastern Fleet to rescue the frigate merely resulted in serious damage to all the rescue ships. The frigate was virtually kept under arrest by the Communists until 30 July, when it managed to slip away in the dark under the command of its new captain, Lieutenant-Commander J. S. Kerans, an assistant naval attaché at the British Embassy in Nanking.[32]

The humiliation of the Royal Navy by the Chinese Communists in the *Amethyst* incident demonstrated the vulnerability of British power in the Far East, particularly in Hong Kong which was the Empire's Far Eastern outpost.[33] More generally, the incident added substance to the view that Hong Kong might 'well become the stage for a trial of strength between Communism and the Western Powers'.[34] It also drew considerable parliamentary and public attention to the possible threat to Hong Kong. As a result, the garrison of Hong Kong was increased from a few thousand under-trained troops to 30,000 soldiers supported by tanks, land-based fighter aircraft, and a powerful naval unit including an aircraft-carrier. The fact that this reinforcement required the calling up of 'substantial numbers' of reservists indicates how seriously the British government had begun to view the security problems of Hong Kong.[35]

In the context of this study, the timing of the *Amethyst* incident and the subsequent massive reinforcement of Hong Kong is particularly important. The incident occurred seven days before Landale gave notice in the Legislative Council of a motion concerning the Young Plan, and it was not over when the Unofficials debated the reform proposals. Moreover, the Cabinet's decision to reinforce Hong Kong on a large scale was not known to the Legislative Councillors until after their 'debate'. Although the Minister of Defence, A. V. Alexander, had announced as early as 5 May, when the House of Commons debated the *Amethyst* incident, that Hong Kong's garrison would be strengthened, the

Cabinet did not decide until 26 May to reinforce the colony substantially. Furthermore, the Cabinet decision was a closely guarded secret because the Cabinet did not wish to commit the British government until the views of other Commonwealth governments could be ascertained. Consequently, when Alexander visited Hong Kong and met the Executive Council on 7 June, he did not refer to the Cabinet decision. Indeed, the Cabinet did not discuss the responses of various Commonwealth countries until 23 June, the day following the Legislative Council 'debate', and the information was not sent to the Governor for his 'personal information' and that of his 'immediate advisers' until 6 July.[36] The Unofficials and the Governor had therefore proposed and discussed the alternative reforms without knowledge of the British government's decision to defend the colony in the event of a Chinese Communist attack.

The Cabinet's decision in the wake of the *Amethyst* incident to defend Hong Kong also meant that the British government had to define more clearly its policy on the future of the colony. After much discussion, the Cabinet finally decided on 29 August 1949 to accept as an 'interim policy' the following joint recommendation of Foreign Secretary Ernest Bevin and Colonial Secretary Creech-Jones:

[W]hile we should be prepared to discuss the future of Hong Kong with a friendly and democratic ['and democratic' was to be deleted under Cabinet instructions] and stable Government of a unified China, the conditions under which such discussions could be undertaken do not exist at present and are unlikely to exist in the foreseeable future. Until conditions change, we intend to remain in Hong Kong, and should so inform other Commonwealth Governments and the United States, while refraining in public from pronouncements which exacerbate our relations with China.[37]

By the time that this recommendation had become policy, Grantham had already submitted to the Secretary of State his recommendations on the Unofficials' alternative reform proposals. This sequence of events suggests that both the Governor and the Unofficials were convinced that the alternative reforms would not be a liability even during a period of political uncertainty.

When the Unofficials' alternative was received by the Colonial Office in late 1949, two of its proposals were strongly disliked: the large Unofficial majority, and the reliance on reserve powers as a safeguard. The Colonial Office officials considered that reserve powers (that is, the reservation of royal assent, the refusal of assent, the disallowance of legislation, and the abrogation of colonial legislation by prerogative legislation) should be used very sparingly and only on matters of prime importance; on matters about which the use of reserve powers could not be justified, the alternative provided no protection. Whilst this criticism is essentially true, the only other option, which was to keep an official majority in the Legislative Council, had few important advantages over the use of reserve powers. As was the case with reserve powers, the official majority could be used only on matters of prime importance. Furthermore, on issues relating to which the use of reserve powers would create hostile comments, the same result would in all probability, though not necessarily, occur if an official majority were to be used. Nevertheless, it is likely that the existence of an official majority would have discouraged the Unofficials from confronting the government, except in matters which they regarded as of fundamental importance. Colonial Office officials considered equally unsatisfactory the alternative's other built-in safeguard, which relied upon the European and nominated unofficial members to counter the Chinese and elected members. The nominated European Councillors could be relied upon to defeat only those motions which demanded that Britain should quit Hong Kong; they could not be relied upon on matters generally styled 'Whitehall versus the Colony'. Whilst this distinction was accurate, it ought to be remembered that (as pointed out in Chapters 2 and 3) the reforms were originally intended to be a bold step to provide a measure of self-government. However, most of the Colonial Office officials involved at this stage were concerned with ruling out any possibility that the government might be defeated in the Legislative Council. Although Creech-Jones did not share this view, he was unable to make an impact. Indeed, for four months following Grantham's recommendation of the alternative, no decision was made, despite strong parliamentary interest. (Between September and December 1949, four direct questions on the Hong Kong reforms were asked in the House of Commons, but no definite answers were given.) Officials in the Colonial Office resorted to delaying tactics, using the fact that Hong Kong was included in the Smaller Territories Committee's study on constitutional development in the smaller colonies as justification for their neglect of the issue.[38]

The specific reasons for the Colonial Office's change of mind towards the idea of reform in Hong Kong are not entirely clear because the relevant 1949 Colonial Office file is still closed. However, the preceding and subsequent minutes indicate that the crucial factor was the coming to power in China of the Communists. Up until that time, there had been no Communist challenge to the Hong Kong government, which both the Governor and the Foreign Office recognized. Yet the Colonial Office, which was dealing with an increasingly serious (and predominantly Chinese) Communist insurgence in Malaya, took a more serious view of the possibility of a similar development in Hong Kong. The advent of what it described as the 'emergency' in Hong Kong subsequent to the *Amethyst* incident and the Communist sweep across south China must have added to the apprehension in the Colonial Office and changed the attitudes of many of its officials. The fact that, with the exception of J. J. Paskin, the officials had not previously been closely involved with the deliberations on the 1945 proposals and the Young Plan undoubtedly made it easier for them to overlook the original intent of the reforms and to choose to mark time. Once they had become concerned about the possibility of the government being defeated in the Legislative Council, however, they deemed the new reform proposals, which provided a meaningful Unofficial majority in matters other than the status of the colony, to be unsafe and, hence, unacceptable.

In January 1950, Creech-Jones decided that despite the so-called 'emergency' in Hong Kong, it was still advisable to proceed with constitutional reform in the colony. He considered Hong Kong to be 'a colony with some political significance' which should not be allowed to fall into the hands of the Communists. He wished to establish British values and standards in the colony, and he disliked the Unofficials' alternative because it departed from the announced policy of the British government. He emphasized that even if this alternative were to be implemented, local government in Hong Kong would remain in an unsatisfactory state; it would bring the government no credit in the face of its parliamentary critics. He therefore deemed it essential that the development of the Urban Council into a municipal council be examined and that the question of redefining the relationship between the Executive and the Legislative Councils be left open. He also found the arithmetic behind the division of seats between the Chinese and the non-Chinese communities bewildering. Out of a population of 1.814 million (of whom 14,000 were non-Chinese), the Chinese were given six seats and the non-Chinese five seats; out of the 16,000 British subjects (of whom 4,000 were Chinese and 12,000 were non-Chinese) who would be qualified to vote, four Chinese and two non-Chinese Councillors were to be elected. In spite of these objections, however, Creech-Jones grudgingly accepted all of the Governor's recommendations. He emphasized that he understood the apprehension of the Colonial Office about the size of the Unofficial majority, but he considered the risk involved to be remote and the provision of reserve powers to be an adequate safeguard. He also finally ruled that the reforms should not be deferred for the consideration of the Smaller Territories Committee.[39]

Officials in the Colonial Office disagreed with Creech-Jones. They thought that he had overestimated the usefulness of the reserve powers and had made an unwise decision. They considered it prudent and practicable to delay the implementation of Creech-Jones' decision at least until after the British general election in February. In order to justify this course of action, they sought and obtained the Governor's concurrence. Creech-Jones did not overrule the officials because he was too preoccupied with the election to spare time for the matter.[40] Subsequent to the election, in which Creech-Jones had lost his seat, it was no longer impossible to put aside his unpopular decision.

After James Griffiths had succeeded Creech-Jones as Secretary of State, the Colonial Office re-examined its policy of constitutional reform in Hong Kong. Two questions were raised: Was the time right? Was the reserve powers clause satisfactory in view of the size of the proposed Unofficial majority? The fact that such questions (particularly the first) were raised indicates that the officials wished not only to avoid Creech-Jones' January decision but also to review the desirability of reform of the Hong Kong constitution. When the Governor was consulted on these matters, he replied that he would like the reform proposals to be handled with 'all reasonable speed'. During the following two months, the Office re-examined the questions of reserve powers and the size of the Unofficial majority. By April, a general understanding had been reached: if the Unofficials' alternative were to be accepted, the size of the proposed Unofficial majority would have to be reduced.[41]

In early May, however, senior officials in the Colonial Office once again questioned the wisdom of proceeding at that time with the reforms. Paskin pointed to the deterioration in Britain's relations with Communist China — not only had Britain failed to establish full diplomatic relations with the Chinese Communist regime, but a complete break of relations was no longer inconceivable. In addition, Deputy Under-Secretary Sir Charles Jeffries considered Hong Kong to be in a state of 'cold siege' and, as such, it should not have any more drastic reform than the granting of a token majority to the Unofficials in the Legislative Council.[42] Paskin and Jeffries had begun to think that delaying tactics should now be replaced by a long-term solution. The solution they favoured would require, however, that Creech-Jones' January decision be set aside, or even that the Unofficials' alternative be turned down. Thomas Cook, the Parliamentary Under-Secretary, agreed with this solution, but Griffiths, the new Secretary of State, did not. Various senior officials subsequently concentrated their efforts on collecting testimonies from official and non-official visitors from Hong Kong to support their case.[43]

In the meantime, the Governor had reviewed his August 1949 recommendations. Although he had not as yet exchanged with the Colonial Office further communications discussing the substance of his 1949 recommendations, he became aware of the attitude of the Office through his Colonial Secretary, J. F. Nicoll, who had extensive discussions with the officials in the Office while he was on leave in England in early 1950. Grantham assumed that in view of Young's pledge on 1 May 1946 to institute constitutional reform, the government was committed to some kind of reform. He still believed that his 1949 recommendations, which were safe enough, represented the best approach, but he also recognized the need to put the mind of the Colonial Office at ease. Consequently, he consulted the Executive Council and modified the Unofficials' alternative in order to ensure the election of a 'safe' Legislative Council in all circumstances. He also proposed that the number of unofficial members in the Urban Council be increased and the number of official members be reduced, but no details were given.[44]

Between June and October 1950, while Grantham was on leave in England, he discussed his revised proposals seven times in the Colonial Office. In the first meeting, he outlined and explained

1 Sir Mark Young (*middle*) restores civil rule to Hong Kong upon his return as Governor (1 May 1946)
Source: *China Mail*

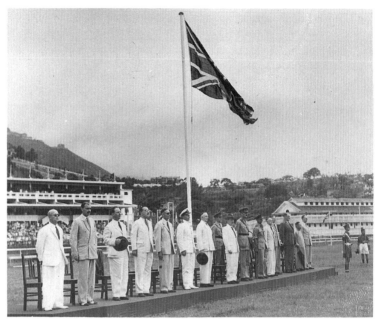

2 Armistice Day parade (November 1946) (*from left to right:* M.K. Lo, D.F. Landale, T.M. Megarry, B.C.K. Hawkins, R.R. Todd, Commodore D.H. Everett, Sir Henry Blackall, Captain Wilson, D.M. MacDougall (Officer Administering the Government), (?) in uniform, (?) in uniform, C.G.S. Follows, Arthur Morse, T.N. Chau, and Bill Williams)
Source: D.M. MacDougall, private collection

3 Sir Alexander Grantham addresses the people of Hong Kong upon his arrival as Governor (King's Theatre, 25 July 1947)
Source: D.M. MacDougall, private collection

4 MacDougall (*seated, second from left*), Grantham (*seated, third from left*), and R.R. Todd, the Secretary for Chinese Affairs (*seated, second from right*), during a lighter moment after a game of tennis
Source: D.M. MacDougall, private collection

5 MacDougall receives the Order of the Brilliant Star from T.W. Kwok on behalf of the Republic of China
Source: D.M. MacDougall, private collection

6 Lord Listowel, Minister of State for the Colonial Office (*right*), is met by D.M.
MacDougall, the Colonial Secretary (*centre*), and A.J.R. Moss, the Director of
Air Services, during Lord Listowel's visit to Hong Kong (March 1948)
Source: *Hong Kong Telegraph*

7 MacDougall retires as Colonial Secretary (1949)
Source: D.M. MacDougall, private collection

外交部
Ministry of Foreign Affairs Janury 26. 1948

My dear Mac. I am deeply touched by your most charming letter.

The honour to which you refer ought not to be for me alone. I believe I am not qualified to say that the major credit for it belongs to you. Without you here, Mac — you write you frankly "without understanding, your mutual judgment, your spirit of co-operation, and your foresight into the future of Sino-British amity, — I shudder to think how we could have passed these months since '45.

外交部
Ministry of Foreign Affairs

I shall always treasure the note of yours for which I send you my heartfelt thanks.

Yrs affly
T.W.

8 A letter from T.W. [Kwok] to Mac[Dougall] following the Kowloon Walled City Incident (January 1948)
Source: D.M. MacDougall, personal papers (published with the permission of Professor D.W.Y. Kwok)

9 General Sir Neil Ritchie (*left*) and Air Marshal Sir Hugh Lloyd (*right*) welcome A.V. Alexander, British Minister of Defence, at Kai Tak airport as Britain strengthens the defence of Hong Kong (June 1949)
Source: *South China Morning Post*

10 The Nathan Road rioters (1 March 1952)
Source: *South China Morning Post*

11 The first post-war Urban Council elections (30 May 1952)
Source: China Mail

12 and 13 The first directly elected post-war Urban Councillors: Brook Antony Bernacchi (*top*) and William Siu Tak Louey (*bottom*)
Source: *South China Morning Post*

his latest idea about the Legislative Council reform, which was to replace direct elections with indirect elections. He insisted that membership of the Legislative Council be restricted to British subjects but he recognized the need to include non-British subjects as members in some of the electoral bodies (namely, the Hong Kong Chinese General Chamber of Commerce and, possibly, the Urban Council). This inclusion was necessary if the amendment was to be presented as an attempt to broaden the franchise, and its real objective, which was to make the Council more easily controllable through the selection of electoral bodies, was to be disguised. Grantham preferred a large Unofficial majority because he thought it to be safe and what the people wanted, but he agreed to reduce its size to a token majority of one if it was politically possible.[45]

Grantham's proposed amendments had an important implication. Unlike the Unofficials' alternative, which provided for direct public participation in elections, Grantham's amendments would merely create an Unofficial majority. The claim that they would introduce 'indirect elections' into the Council was illusory. Although Hong Kong technically only had appointed members in its Legislative Council, two of its unofficial members had by convention been appointed by the Governor on the recommendations of two (predominantly expatriate-British) 'representative bodies' — the Unofficial Justices of the Peace and the Hong Kong General Chamber of Commerce. Although they could be regarded as nominated members, they were more suitably termed indirectly elected members since this was how they had been chosen since 1884. Grantham's latest amendment amounted merely to the formal institutionalization of this practice of nomination and the acceptance of a few 'representative bodies', which were not primarily expatriate-British in their membership, as nominating bodies. Young had proposed something similar in his supplementary proposal regarding the Legislative Council. Indeed, Grantham's revised suggestion regarding the Legislative Council would have been virtually identical to Young's supplementary proposal if the size of the Unofficial majority had been reduced to one.

Whilst officials in the Office generally welcomed Grantham's revised proposals as being safer than the Unofficials' alternative, they were concerned that they might be unacceptable to the general public in the colony. Grantham assured them, however, that his latest idea had been 'well received' when he had mentioned it informally to the local English-language newspaper editors. He expected only the Chinese Reform Association, which he regarded as left-wing, to protest. The Chinese-language press and the Hong Kong Reform Club, which he omitted to mention, were either considered to be unimportant or unlikely to raise important objections. There are indications that the Governor did not think highly of the Reform Club because it was already restless about the long delay. It had in fact telegraphed the Prime Minister in May to request that the British government pay early attention to the Hong Kong reforms. In the course of his discussions, Grantham also mentioned his supplementary proposal to change the Urban Council. He explained that no Urban Council elections had been held after the war because the Young Plan had been put forward. He added that elections to the Urban Council could be carried out 'quickly' and 'without difficulty'. The details of the Urban Council reform were not discussed, however, because it was not considered to be an important or controversial issue. After two meetings between Grantham and the officials, Grantham's latest proposals were submitted to the Secretary of State.[46]

Griffiths disliked the proposals because they were 'retrogressive' and contradicted the policy of the government. He said that Malcolm MacDonald, the British Commissioner-General in South-east Asia, had recently warned him that Britain must plan quickly for the future of its East Asian colonies because they would be the only ones in the region without self-government; all other South-east Asian countries with a colonial past would shortly have secured it. Griffiths also emphasized that the only way to retain Hong Kong 'was to make the population want to stay within the Empire'. He preferred the Young Plan: since Hong Kong was growing into a city-state and could not afford two main authorities, it would be wise to establish a municipal council as the vehicle for constitutional advance.[47]

Paskin and Grantham dissented from Griffiths' view. They stressed that Hong Kong was a very special case and that it should not be compared with Malaya. They argued that Hong Kong's future lay either as a British colony or as part of China, and that the only way in which it could be kept British was by making its continued existence valuable to the local people, to China, and to the world. They thought that Hong Kong could not become an independent city-state. Although Griffiths was not convinced by Paskin and Grantham, he did not have adequate knowledge of the region to counter their arguments. He thought that the officials simply 'did not want to do anything in Hong Kong' and that they would probably refer to the Far Eastern situation, which had been deteriorating since the outbreak of the Korean War, as justification. He therefore preferred to postpone a decision on the Hong Kong reforms for six months until the Far Eastern situation became clearer.[48] It is noteworthy that, with the exception of Griffiths, no one in the Office had linked the Korean War with the Hong Kong reforms at that time.

While Griffiths was correct in assuming that the Colonial Office and Grantham did not wish to reform the Hong Kong constitution, it does not mean that they did not intend to do something. Although they contemplated making a public reply to the Governor's recommendations of August 1949, this did not mean that they were in a hurry to introduce any reform. They were determined to 'drag the correspondence [between Whitehall and Hong Kong] out for a matter of eighteen months' before introducing any reform. They were in a dilemma: they wished to abandon the Unofficials' alternative and to replace it with Grantham's revised proposals, but Griffiths would approve only the Unofficials' alternative (or the Young Plan) and not Grantham's revised proposals. They wished to convince the people of Hong Kong that they had been doing something about the reforms, thus preventing the local leftists from making political capital out of the delay. Indeed, the long delay had caused some concern in the colony. The Hong Kong Chinese Reform Association had sent a letter to Sir Hilton Poynton (Deputy Under-Secretary) in July to remind the Colonial Office that the Governor had promised a reply within three months when the Association, together with 139 other Chinese organizations, had petitioned the government for reform a year ago.[49] Consequently, they tried to convince Griffiths to send a non-committal reply to Hong Kong.[50]

In order to make Grantham's latest proposals more acceptable to Griffiths, the Colonial Office and Grantham amended the proposals slightly. They now placed more emphasis on building up the Urban Council, and agreed to expand the Council's franchise, which was based on the jury list, to include British subjects hitherto ineligible because of an inability to speak and write in the English language. Grantham added that while he had planned to give the new Urban Council the right to nominate only two members to the proposed Legislative Council, he would now agree to its nominating three. Despite these changes, Griffiths refused to sign any dispatch which would put him in a position of suggesting that Hong Kong should abandon direct elections in favour of indirect elections, which he considered to be retrogressive.[51]

Paskin and Grantham subsequently attempted to persuade the unofficial members of the Legislative Council in Hong Kong to put forward Grantham's revised proposals as their own and, thus, save Griffiths from having to suggest them. J. F. Nicoll, the Officer Administering the Government in Hong Kong, was told of the conclusions reached in the Colonial Office and was asked to consult his Executive Council. According to Nicoll, he consulted the Council on 5 September, but the Executive Council minute of that date has no record of this issue being discussed. For reasons of their own, the Unofficials (or, less likely, Nicoll) did not wish their views on the matter to be recorded. In a 'secret and personal' letter to Paskin, Nicoll said that the Unofficials hoped that they would not be asked to propose something 'less liberal than the 1949 proposals'. However, they only accepted the representation of the non-British Chinese in the proposed Legislative Council as a temporary expedient — they hoped to 'revert' the franchise to British subjects at a later stage. They wanted an additional unofficial member to be appointed by the Governor in the proposed Legislative Council, and they preferred not to widen the franchise of the Urban Council. The Unofficials were asking, therefore, for the best of both worlds: they sought to put further restrictions on Grantham's revised proposals without appearing to have suggested them publicly.[52]

Grantham discussed the Executive Council's views with the Colonial Office before he left England for Hong Kong. The Office and Grantham found the Executive Council's objection to expansion of the Urban Council franchise to be unreasonable. They were opposed to the increase in the number of appointed unofficial members by one because Griffiths would not have agreed to it. They thought that if there were to be an additional unofficial member, he would have to be nominated either by the Social Welfare Association or by the Urban Council. They rejected the suggestion that the latest proposals be treated as a temporary measure since, as a matter of principle, no constitutional reform should ever be so regarded. They also insisted that the Unofficials should put forward the newly revised plan. Grantham and the Office were now determined to carry out the understanding they had reached in London.[53]

The proposals were finalized after Grantham's return to Hong Kong in October. He held further consultations with the Executive Council and came to the following conclusions. The additional unofficial member of the Legislative Council should be appointed by the Governor (the Social Welfare Association being regarded as inappropriate as a nominating body). The Unofficials were also strongly opposed to the Urban Council's nomination of a total of three members to the Legislative Council lest it 'give the subordinate body a disproportionate weight in the deliberations of the Legislature'. Grantham further decided that the Urban Council should be reformed in two stages. The Council should first be reconstituted with direct elections reintroduced without changing the franchise. The Urban Council franchise should then be extended to include those people who would have been eligible to vote in the proposed Legislative Council elections under the Unofficials' alternative. The Colonial Office supported Grantham's recommendations.[54]

The proposals finally agreed at the end of 1950 were that both the Legislative and Urban Councils should be reformed. The new Legislative Council would have sixteen members, including the Governor who would preside and have a casting (but not an original) vote. Of the fifteen members, four would be official, five would be appointed unofficial, and six would be indirectly elected members. Of the indirectly elected members, two (one Chinese and one non-Chinese) would be nominated by the Unofficial Justices of the Peace, two would be nominated by the reconstituted Urban Council, and one each would be nominated by the (predominantly British) Hong Kong General Chamber of Commerce and the (predominantly Chinese) Hong Kong Chinese General Chamber of Commerce. According to the Colonial Office's own reckoning, of the total of sixteen votes, at least ten would be safe votes (namely, all four votes of the officials, three of the five votes of the nominated members, one of the two votes of the Justices of the Peace, the Hong Kong General Chamber of Commerce vote, and the Governor's casting vote). The Urban Council would be reconstituted as follows: the number of ex-officio members would be reduced from four to two; the number of directly elected members would be increased from two to four; and the number of appointed members would remain unchanged at six. An official would continue to chair the Council. The size of the new Urban Council would therefore remain at thirteen members, but it would have an unofficial majority of seven instead of three. However, the duties and power of the Urban Council would remain unchanged — a clear indication that all changes to this hybrid of a sanitary department and an advisory board were primarily cosmetic. The changes in the two Councils were also to be co-ordinated. After the Urban Council had been reconstituted on the basis of the old franchise, it would nominate two of its unofficial members to sit temporarily in the Legislative Council. After the franchise of the Urban Council had been amended, new elections would be held for the Urban Council, and the new Urban Council would then elect its two representatives to the new Legislative Council.[55]

The Colonial Office had not yet secured Griffiths' approval of the above reforms but it was confident that approval would be forthcoming if other interested departments also supported them at the official level. Consequently, it proceeded to consult the Foreign Office, the Commonwealth Relations Office, the Ministry of Defence, and the Treasury. With the exception of the Foreign Office, none of these departments had any important objection.[56]

The Foreign Office disliked the latest scheme because it would provide excellent grounds for the Chinese Communist propagandists to attack the British authorities in Hong Kong. As one of its officials described it, the reform proposals were 'undemocratic', and could be criticized by the Communists as an 'instance of the hypocritical insincerity of the imperialist oppressors' aiming at 'brutally crushing the rightful interests of the Chinese in the Colony'.[57] In its reply to the Colonial Office, the Foreign Office stressed that it was undesirable to take an action which would certainly provoke the Chinese Communists — who had remained silent about the British possession of Hong Kong — into launching a propaganda attack at a time when the Far Eastern situation was 'particularly serious'.[58] Although a policy regarding Hong Kong's future status had already been made, the Foreign Office was reluctant to take any risk of stirring up a hornet's nest; it objected to the publication of any dispatch on constitutional reform in Hong Kong.

Paskin in the Colonial Office recognized that Griffiths, who had expressed concern in the summer of 1950 over the Far Eastern situation, would use the Foreign Office argument to turn down the revised proposals. He passed the matter to Grantham virtually for a decision. Others in the Colonial Office were more concerned about the possibility of parliamentary questions if there were further delays in the Hong Kong reforms, but they thought that the delay could be justified by reference to the general instability of the Far East caused by the Korean War.[59]

Of particular importance is not the fact that the Colonial Office had begun at the beginning of 1951 to relate the reforms to the Korean War, but that it had not done so earlier. In fact, the war had broken out during the first week in which Grantham had discussed his revised proposals in the Colonial Office (in June 1950). Yet, on record, it was Griffiths who had indirectly linked the reforms to the war; Grantham and the officials had not done so, even though as early as July 1950 the Colonial Office had collaborated with the Foreign Office and the Chiefs of Staff Committee in examining the impact the Korean War would have on Hong Kong. At that time, they had found no evidence that the Chinese Communists intended to invade the colony, but they also recognized that the Communists had increased their armed strength near the Hong Kong border to over 200,000 men and that they could attack the colony with a week's notice. In August, it was noted that if Britain and Communist China were to break off relations, the Communists might invade the colony. By the middle of October, Chinese Communist forces had crossed the Yalu river and entered the Korean War; British soldiers sent from Hong Kong were confronting Chinese Communist forces in Korea. As a result, the Colonial Office had to prepare for the possible spill-over of the 'international police action' in Korea into Hong Kong.[60] Nevertheless, the Colonial Office had ignored the situation in Korea when it finalized the reform proposals with Grantham, indicating that it had not hitherto thought that the crisis in Korea was directly relevant to the reforms in Hong Kong, but that it was now ready to accept the Foreign Office's suggestion subject only to the Governor's view.

Grantham was pleased to learn of the Foreign Office's objection. He said that the Executive Council generally shared his view. The Chinese Councillors took exception to one point, however. They 'expressed anxiety lest the fear of provoking retrocession demands should be taken as a pretext for indefinitely deferring the proposals'. In the light of the Chinese Councillors' previous lack of enthusiasm for reform and the strong private criticism of the British government for not making an unequivocal statement about Hong Kong's political status, this expression of anxiety seems insincere and may have been another opportunistic manoeuvre to show that they were not reactionaries. Grantham said that as long as the local people were busy struggling to survive the first post-war depression (which had been caused by the imposition of American trade restrictions against Hong Kong as a result of the Korean War) there would be no objection to constitutional reform being delayed. Indeed, he proposed 'to say no more' about the matter until the Far Eastern situation had become settled.[61] The reforms were consequently put aside.

Between early 1949 and 1951, the whole idea of reform of the Hong Kong constitution had undergone a subtle and gradual but fundamental change. Not only had the original reform proposals twice been replaced by different sets of proposals, but the principle governing the whole exercise had been altered. The rationale behind the reforms at the beginning of 1949 had been to give the inhabitants of the colony greater local self-government. By the summer of 1950, however, it had become one of apparently making good the British government's promise by introducing some kind of reform which would not in any way lessen Whitehall's control over the colony. It had been the Colonial Office which had changed its views most obviously. Prior to 1949, various officials in the Office had had diverse views on how to reform the constitution, but they had generally accepted the pledge of 1 May 1946. From the autumn of 1949 onwards, the Colonial Office gradually reached a consensus that the 1946 pledge was a self-made trap, and its only concern was to seek a practical way of getting out of it. The only people inside the British government who were still committed to fulfilling the promise of 1946 were the two Labour Secretaries of State, particularly Creech-Jones. This state of affairs explains why the deliberations of 1950 were almost entirely concerned with how to make the latest but retrogressive proposals acceptable to Griffiths. Paradoxically, Governor Grantham, who was the moving force behind the transformation which took place during this period, had not changed his mind. From the very beginning, he had found the 1946 pledge a matter to be 'regretted' because he thought that major reform was inappropriate for Hong Kong.[62] At the end, he was pleased for the same reason to be able to put aside those reform proposals he had personally drafted and forcefully defended. Grantham had not, however, rejected any of the reform proposals merely for the sake of doing so; he simply thought that he knew what was best for Hong Kong, and that that excluded democratic reform. He was not opposed to minor changes, having quietly expanded both the Executive and Legislative Councils, albeit only within the framework of the existing constitution. (In 1948, he increased the size of the Executive Council from nine to twelve by adding an official member, an expatriate-British member, and a Chinese-British unofficial member. In 1951, he also expanded the Legislative Council by appointing another official member and an additional Chinese-British unofficial member.[63]) In the final analysis, despite the actual changes which Grantham introduced, the transformation of the principle and content of the Hong Kong reforms during this two-year period was nothing less than a metamorphosis.

The Smaller Territories Committee

On 25 October 1949, less than two months after Grantham had recommended the Unofficials' alternative to the Colonial Office, the Secretary of State inaugurated the Committee of Enquiry into Constitutional Development in the Smaller Territories (the Smaller Territories Committee). This Committee was set up to 'enquire into the present constitutional position of the smaller Colonial Territories and the probable trend of their future political development' and to make recommendations in the light of their findings. More specifically, it was to answer three questions: Should a common policy or general principles governing a common policy on constitutional development in all smaller colonial territories be laid down, or should particular policies be defined for particular territories? How should the colonial peoples of the smaller territories be led towards developing a sense of devotion to and partnership in the Commonwealth? And what was the most efficient and economic way of reforming the political, legal, and administrative structures of these territories? The Committee was chaired by Sir Frederick Rees, Principal of University College, Cardiff. Its five non-official members were Margery Perham, Reader in Colonial Administration, Oxford;

Vincent T. Harlow, Beit Professor in Commonwealth History, Oxford; Sir John Maude, Deputy Chairman of the Local Government Boundary Commission; T. Reid, Labour Member of Parliament for Swindon and formerly in the Ceylon Civil Service; and A. D. Dodds-Parker, Conservative Member of Parliament for Banbury and previously in the Sudan Political Service. The Colonial Office representatives were Sir Charles Jeffries, Deputy Under-Secretary, and Sir Kenneth Roberts-Wray, Legal Adviser to the Secretary of State. Hong Kong was included in the scope of the Committee's inquiry because it was considered to be one of the twenty-three smaller territories which for strategic reasons or because of an insufficiency in resources, or both, could not achieve full self-government.[64]

The original mover behind this idea to form a committee of inquiry was C. R. Attlee, the Prime Minister, who was dissatisfied with the way in which constitutional reform in Gibraltar was being handled. Attlee observed that until now no comprehensive study of the constitutional problems in the smaller territories had been made; the ultimate objective for these colonies had never been defined; and it had been too readily assumed that the Westminster model could be reproduced in miniature even in the smallest colonies. It was consequently decided that a special committee with both official and non-official members should be set up to study the matter confidentially, lest false expectations or even political agitation be provoked in the colonies concerned. The Committee was not appointed to tackle the immediate constitutional problems of the smaller colonies, but to study their problems and to make long-term recommendations. The British government had recognized that it had to do something to retain the loyalty and assuage the feelings of the smaller colonies, which would increasingly aspire to full responsible self-government — a goal meant not for them, but for the larger colonies which could ensure their own people 'both a fair standard of living and freedom from oppression from any quarter'.[65]

The Committee made an interim report to the Secretary of State in March and submitted its full report in August 1951. Its final report was circulated for comment to all Assistant Under-Secretaries in the Colonial Office and to some Governors who had special knowledge of the subject. The report and the comments received, which were mostly unfavourable, were submitted to the new Conservative Secretary of State, Oliver Lyttelton, in early 1952. Lyttelton decided that although some of the recommendations of the Committee might be adopted in suitable cases, the report should be put aside. Indeed, the report was not circulated to the Cabinet even though a copy was sent to the Secretary to the Cabinet.[66]

The Committee began to pay attention in December 1949 to the constitutional problem of Hong Kong. It thought that its city-state concept for Gibraltar and Aden, which combined municipal and legislative functions in one single council and provided the unofficial members of that council with certain executive duties, might also be applicable to Hong Kong.[67] It did not closely examine the issues involved, however, until the end of March 1950.

The Committee recognized that the fundamental problem in Hong Kong was how to give the inhabitants, who were overwhelmingly Chinese, greater local self-government without endangering the security of the colony, which was of great strategic and commercial importance. The Committee believed that a new constitution would soon be required for Hong Kong because the government had already committed itself, and it would be desirable to meet the local aspirations before discontent grew. It also realized, however, that the colony's strategic and commercial value to Britain made it unwise to let the colony 'aspire to dominion status, which would leave it free to quit the Empire and join the enemies of liberty' in Communist China. The fact that most of the local people had retained their Chinese nationality indicated to the Committee that they could not be relied upon to vote to keep the territory British. Consequently, any system based on a wide popular franchise which included non-British subjects (such as the Young Plan) would be objectionable. The Committee also thought that, for purely administrative reasons, Hong Kong should not be over-burdened by two sets of elections, one at the municipal level and the other at the 'parliamentary' level. The existing Urban Council should therefore be abolished if politically possible. The Committee found the Unofficials' alternative attractive, but rejected it on the grounds that it left the way open for the appointment of an *ad hoc* commission to examine the possibility of reforming the Urban Council. Such a course was considered to be dangerous because, in view of the size and importance of Hong Kong as a city, an *ad hoc* commission of experts could hardly be expected to recommend anything less than that the Council be given full municipal status and all that that would entail. It was not until July that the Committee agreed on its recommendations for Hong Kong.[68]

The Committee proposed that the colony be turned into 'The City State of Hong Kong'. It would have a Governor who would enjoy all of the customary prerogative powers, executive duties, and reserve powers. The existing Executive Council would be replaced by a Privy Council, with the Governor as President, and the State Secretary (that is, the old Colonial Secretary), the Attorney-General, the Commander of the British Forces, the Financial Secretary, and the Secretary for Chinese Affairs as ex-officio members. In addition, it would also have members nominated by the Governor and appointed by the Secretary of State. The role of the Privy Council would be purely advisory; the Governor would be free to accept or reject the Council's advice. The existing Legislative and Urban Councils would also be abolished and replaced by a State Council, which would have both legislative and executive functions. The State Council would be constituted by four ex-officio members (the State Secretary, the Attorney-General, the Financial Secretary, and the Secretary for Chinese Affairs); two representatives elected by the Hong Kong General Chamber of Commerce; two members elected by the Hong Kong Chinese Chamber of Commerce; and seven other members directly elected on a single-member constituency basis. For legislative purposes, the Council would appoint four standing committees, each chaired by one of the ex-officio members. In the event of a deadlock or a disagreement between the Chairman and the majority of the members in a committee, the matter would be referred to the whole house. The franchise for the seven elected seats would include all adults who could prove continuous residence in the colony since 1946 and who could pass a simple literacy test in the English or Chinese language — by implication, this requirement included the non-British Chinese residents.[69]

The Committee discussed its city-state model with Grantham while he was on leave in England during the summer of 1950. Grantham opposed the model on the grounds that the people of Hong Kong would 'certainly' misinterpret the scheme to mean 'the first stage in a British withdrawal'. He preferred that his own revised proposals, then under consideration in the Colonial Office, be adopted. The Committee pointed out that his approach, which 'left the election of a legislature in the hands of representatives of the richer classes' was 'undemocratic'. The Committee added that it would 'weaken the British position both in regard to any Chinese Government's claim to the incorporation of Hong Kong in[to] China and in regard to world, and particularly American, opinion'. Grantham replied that any Chinese government, whether Communist or Nationalist, would press for the return of Hong Kong, however democratic and useful (to China) the colony might be. He argued that it was also virtually impossible to evoke a sense either of local loyalty or of belonging to the British Commonwealth because Hong Kong could not ban the free movement of people between the colony and China without either jeopardizing her entrepôt trade or antagonizing China. He asserted that the introduction of direct elections into the colony would 'result in the dominance in Hong Kong of Chinese politics', notwithstanding his private assurances during the preceding few weeks to the Colonial Office that his 1949 recommendations for direct elections were still generally safe — the only danger under the 1949 proposals was that some directly elected Chinese Unofficials might speak against and embarrass the government if some of their motions were defeated. More importantly, in his own estimation, there was 'no real demand for such a step' locally. However, he stressed that the Young Plan had aroused public expectations and that it would be necessary to do something to satisfy 'the more responsible people' in the colony. He added that the oligarchical nature of his proposals would be accepted locally because the people were used to that conception. In view of Grantham's evidence, the Committee doubted whether it could make any useful recommendations for Hong Kong, which was bound up with foreign policy and unique in its problems. In its final report to the Secretary of State, submitted in August 1951, the Committee stressed that although it thought that the Governor's proposals were inappropriate for the colony and it preferred the city-state model, it thought that the city-state model was not practicable in view of the 'exceptional circumstances' existing. The Committee consequently decided not to recommend any model for Hong Kong.[70]

The Committee made no impact on the Hong Kong reforms for

three reasons. It had been appointed to examine long-term, not immediate, policy. Secondly, its final report was not accepted. Finally, its lack of a real understanding of the colony had prevented it from making any useful recommendations. Grantham's reference to the special circumstances in Hong Kong in support of his own proposals gave the Committee no choice but to accept his arguments, knowing too little of the local situation to question his assertions. Whilst Grantham's observations were largely valid, he unwittingly misled the Committee simply by the fact that his statements went unchallenged. A case in point was his assertion that direct elections would result in the dominance of Chinese politics in Hong Kong. While this might be true of direct elections based on universal suffrage, it would not have been the case if the franchise had been restricted to British subjects. Grantham was himself aware of this distinction when he had assured the Colonial Office that the Unofficials' alternative was safe. Although he was probably prepared to elucidate upon his assertion, he was never called upon to do so.

Security and Local Politics

As examined earlier, the Colonial Office looked at the security problems of Hong Kong in 1949 with some apprehension, but the Governor generally took a more relaxed view. This was due to the differing assessments of the Governor and the Colonial Office of the importance of the rise of the Communists in China. Some officials in the Colonial Office were almost hypersensitive to the fact that the faction winning the civil war in China was Communist.[71] The Governor was not so concerned, not because he had any illusion that the Chinese Communists were 'radish Communists' (that is, red on the outside, but white on the inside) or 'Titoists', but because he looked at the matter from a different angle. From Grantham's point of view, of real importance to Hong Kong was the fact not that the Chinese Communists were orthodox Communists, but that they were 'well-disciplined and dedicated' people who would in due course set up a strong government in China, and that a strong Chinese government would demand the return of Hong Kong.[72]

As far as the immediate situation was concerned, Grantham recognized that the Chinese Communists could destabilize the colony, but he was also aware that they had so far maintained a 'scrupulously correct attitude' towards the colonial government. In spite of this, he strongly advised the British government publicly to declare that it was 'determined to maintain its position in Hong Kong, come what may'. His main object in so doing was to reassure the business community of the colony. As to the prevention of Communist hostility against the colony, he considered that the best approach was to show determination for its defence, which could be achieved by an unequivocal statement by the British government and by strengthening the local garrison.[73]

As a general security measure, the government of Hong Kong legislated to prohibit all societies which were branches of, or affiliated with, the Triad or foreign political parties, including both the Chinese Communist Party and the Kuomintang. Other legislation was also enacted to ban politically inspired strikes, to register the entire population of the colony, to deport politically undesirable aliens, and to close Communist-controlled schools (the most important of which, the Tat Tak College, had been closed down at the beginning of the year). By the end of 1949, the necessary legislation was largely completed after the Secretary of State approved the Emergency (Principal) Regulations, which could be brought into operation piece by piece by the Governor-in-Council and which would give the Governor-in-Council almost dictatorial power if fully enforced. The government had also increased the size of the Police Force to double that of 1941, re-established the local defence force and the essential services, stockpiled essential food supplies for a period of six months, and obtained the support of the Army and Navy in closing both the land and sea borders of the colony if a major influx of refugees were to occur. Hong Kong had thus prepared itself for any exigency. Indeed, the local Commissioner of Police felt confident enough to confirm to the British Minister of Defence as early as June that his force could clamp down on all subversive elements at any moment if the situation became difficult.[74]

The approach of the government of Hong Kong was to watch the potentially subversive elements closely, while ensuring that it had the power to deport quietly those individuals who had clearly broken the law. It was thought that a general suppression of the Chinese Communists before they had begun to make trouble would hamper the prospect of trade between Hong Kong and

Communist China. It would also drive the Communists under-ground without affecting whatever plans they might have for Hong Kong, and might even provoke them to turn to violence. The rationale was that 'a sudden emergency situation of some particular kind' could arise in Hong Kong 'at any time without the occurrence of a *general* emergency'. The government's guid-ing principle was that it should neither over-react nor appear soft to the potential Chinese Communist challenge. It intended to strike at the Communists only at the time and circumstances of its choosing.[75]

The first occasion when Grantham supported a hard line against the Chinese Communists in the colony was the *Amethyst* incident. On 22 April 1949, the day after the Royal Navy's final attempt to save the frigate had ended in failure, the Executive Council proposed to suppress the Communist newspapers and the Hsin Hua News Agency because of their 'steadily worsening tone . . . particularly in respect of comments on the Atlantic Pact'. The real reason for the proposal, however, was the *Amethyst* incident.[76] Both the Chinese and British Councillors found the *Amethyst* incident to be 'painfully reminiscent' of the loss of HMS *Prince of Wales* and HMS *Repulse* during the last war. The *Amethyst* incident was particularly disturbing because the British government had refused to make an unequivocal statement about the future of the colony; the Councillors wanted a show of British determination. One prominent Chinese-British resident frankly told a British Member of Parliament that people like himself wanted Hong Kong to remain British, but that if Britain aban-doned them to the Communists, they would have to make their peace with the other side. In such circumstances, and particularly with an eye on public morale, Grantham could hardly have done anything other than relay the Councillors' request to the Secret-ary of State. Nevertheless, he did not pursue the issue when British diplomats in China sounded a note of caution. He saw eye to eye with the British Ambassador that 'precipitated action would aggravate relations' between Hong Kong and Communist China. Grantham did not think that the propaganda was a chal-lenge to the government; he merely did what was necessary in order to console the Councillors.[77]

When Grantham was confronted with what he believed to be the first Chinese Communist challenge, he was determined to respond with strength. The occasion was the Tramways strike in the winter of 1949. Grantham thought that the strike was part of a co-ordinated attempt by the left-wing Hong Kong Federation of Trade Unions to make a bid for control over the local labour movement by March 1950. By his account, the Federation drew up its plan in May 1949, launched it in September, and then called for strikes by five public utility unions: the Telephone, Tramways, Gas, Hong Kong Electricity, and China Light and Power (that is, Kowloon electricity) Unions. The Federation was said to have instructed the Telephone Union and the electricity unions to make demands for a special allowance and some subsidiary benefits and, when these were rejected, to join in a co-ordinated strike. However, none of these unions chose to strike. As a result, the Federation turned to the Tramways Union, which was relatively the most heavily infiltrated by local leftists, to act as its spearhead. In December, the Union demanded an additional special allowance of three dollars a day. The company rejected the demand. The Tramways Union then presented the company with an ultimatum, which was ignored. Consequently, the Union decided to stop collecting fares in trams from Christmas Eve onwards. The company retaliated after Christmas by sacking all conductors and closing the company's premises to all Union members. The government, which to all appearances was not involved, in fact stood behind the company to test the strength of the Union.[78]

The Governor, who looked on the strike as a challenge to his government and did not consider the tramways to be an 'essential service', decided to make use of this occasion to break the back of the Federation. His strategy was to allow the strike to 'take [its] natural course ... so that the political planning and leadership of the Federation would be brought more and more into the open and that, ultimately, the Federation would be discredited not by an isolated incident but steadily and thoroughly over a period'. The opportunity to deal the Union a major blow arose in late January 1950. On 28 January, a Tramways Union leader introduced political issues into his speech. The police then warned the Union not to conduct any further meetings in its Russell Street premises. On 31 January, the Union evaded the police ban and relayed speeches made inside its Russell Street office to the street using a loudspeaker. The police removed the loudspeaker and a riot followed. The police, who were obviously very well prepared, cleared the 5–6,000 people in the street in just over an hour. Subsequently, several leaders of the strike, including the Chairman of the Tramways Union, the chief picket, and an 'adviser' from Canton, were deported. Within a fortnight, the new leaders of the Union gave up the struggle and returned to work on the condition that dismissed workers be reinstated; special allowances and other demands of the Union were left for later discussion. Following the end of the strike, a deportation order was also made against Cheung Chumnam, Chairman of the Hong Kong Federation of Trade Unions, who had been absent from the colony prior to the Russell Street incident. The government had thus been able to achieve its objective: it had not only triumphed over the Tramways Union, but had also crippled the Federation for at least a few years.[79]

The way in which the government of Hong Kong handled the Tramways strike and its aftermath exemplified its attitude towards the labour problem in the colony. The government had been biased in favour of the company from the very beginning — hardly surprising, given Grantham's view of the strike and the fact that two of the Unofficials, M. K. Lo and P. S. Cassidy, were also directors of the Tramways Company. In addition, the Commissioner of Labour regarded the Union's demand as unjustifiable because it virtually called for a thirty-three per cent increase in pay, while the rise in the cost of living was only twelve per cent. The demand of the workers was, however, not as unreasonable as the Commissioner maintained. For almost five years since the end of the war, all pay increases to the workers had been adjustments in the cost of living allowance which was calculated according to the price of food. The standard of living of the workers was therefore still kept at subsistence level, whereas the company was increasing its profits every year. If the workers wished to improve their lot, they had to demand an increase higher than the rise in the cost of living index, and there is no indication that their demand was higher than the company could afford. The Commissioner of Labour had failed to recognize that the workers did have a genuine economic grievance — a point on which almost all the leading local Chinese-language newspapers of whatever political persuasion agreed.[80]

Although some leaders of the Tramways Union (particularly those closely connected with the Hong Kong Federation of Trade Unions) probably did have political considerations, most of the workers regarded the strike as primarily, if not solely, an industrial action. If the majority of the strikers had been politically minded, they would not have given up the struggle without making a stand once the police had been sent in. The strikes of 1925–6, which did have political objectives, provide a good contrast. The post-war experience of riots in the Walled City of Kowloon and in Shameen further suggest that if the 5–6,000 people in Russell Street on 31 January had been in a confrontational mood, the police could not have cleared the area in just over an hour without causing at least one fatality, however well prepared they might have been. The workers supported the Union because its demands were attractive to them; they abandoned the hard-liners once they knew that they could not deliver their promises.[81]

If the strike demonstrated anything, it was how little influence the Hong Kong Federation of Trade Unions had over its affiliated unions before, during, and after the strike. If it had in fact designated five of its affiliated utility unions for a trial of strength against the government, it had failed from the beginning since it could only mobilize one to take action. The inability of its most heavily infiltrated union (the Tramways Union) to carry on the struggle after it had suffered an initial setback at the hands of the police on 31 January further exposed how little control it had over the workers. After the strike had ended, the Federation not only failed to fight back but also entered a period of decline. Its prestige had been shattered and its membership had fallen, so much so that by June, one of its affiliated unions had ceased to exist and another saw its membership drop from 180 to 6.[82]

The strike was in reality neither organized nor actively supported by the Chinese Communists. It occurred because the leaders of the Tramways Union (and probably also the Hong Kong Federation of Trade Unions) miscalculated the support they would receive from the Communist authorities in China, and because the government (as well as the company) was prepared for confrontation. The Chinese Communist authorities in Canton considered the strike to be 'mistimed and badly organized'. When the local leaders of the strike went to Canton to seek financial and other backing, they were given only one 'adviser', whose presence in the colony was kept secret — a token gesture rather than substantive support. This lack of real support confirmed the British Cabinet's scepticism about the claim that 'Communist intervention was solely responsible for the incident'. In fact, after the strike was over, even the Governor admitted that 'the trouble stemmed from local causes rather than from outside instigation'. Throughout the incident, neither the Hong Kong Federation of Trade Unions nor any other Chinese Communist organization made a stand against the Hong Kong government; the whole affair was left to appear as an industrial action in which the government ostensibly played no important role.[83]

The Chinese Communists had, in fact, not changed their fundamental attitude towards the British and Hong Kong, which had been indirectly expressed to the British authorities by Chiao Mu at the end of 1948.[84] Their main objective remained the winning of the civil war in China, not the subversion of Hong Kong. For this reason, the leading Chinese Communists in Hong Kong, including Chiao Mu and Fang Fang, left the colony for duties inside China in the autumn of 1949 as the Communist forces took over south China. Their departure was followed by 'a general exodus of prominent Communists from Hong Kong' after the Communists occupied Canton. Even *Hwa shiang pao*, hitherto the most important Communist newspaper in the colony, was transplanted to Canton to become the Party's mouthpiece, the *Nan fang jih pao*.[85] Communist documents captured by the Hong Kong government provided further evidence of the role played by Hong Kong in the civil war in China. Documents seized in a raid on Fang Fang's home in the summer of 1949, for example, confirmed that the Communists regarded the British colony as a vital base for their struggle inside China. Following the Hong Kong government's introduction in 1949 of the Societies Ordinance, which the Communists considered to be both 'aggressive and provocative', they continued to oppose 'leftist deviation of a venturesome nature'. They stressed the need to 'extensively and resolutely make use of the open and legal tactics and exert [their] best to struggle for registration', even though they also cautioned against an 'over-tendency of abiding by the law'. The Chinese Communists intended to fight back only if they were attacked, and they hoped to do so in a legal and constitutional way.[86]

The disputes concerning Chinese state properties in Hong Kong, particularly the Chinese aircraft grounded in the colony, exemplified this approach to the British authorities. The aircraft concerned originally belonged to the Chinese National Aviation Corporation, a corporation largely owned by the Kuo-

mintang government, and the Central Air Transport Corpora-
tion, a department of the Kuomintang government. In the sum-
mer of 1949, the two airlines moved their main bases to Hong
Kong as a result of the Communist occupation of Shanghai. In
November 1949, following the inauguration of the Communist
People's Republic of China, an effective majority of the em-
ployees of both corporations declared themselves in favour of the
Communist government; twelve planes then left Hong Kong for
Communist China. Subsequently, the pro-Communist and pro-
Kuomintang employees of both corporations applied injunctions
against each other and the remaining seventy-one planes were
grounded in the colony. In December, the Kuomintang and the
Americans, who feared that when Britain recognized Communist
China the aircraft would be handed over to the Communists,
hastily established an American company, Central Air Transport
Incorporated, to which the titles of all the aircraft were obstens-
ibly transferred. In February 1950, the Hong Kong Supreme
Court duly awarded all the aircraft to the Communist govern-
ment, Britain having recognized Communist China on 6 January.
The United States then applied very strong pressure to Britain to
disregard the legal position and deny the aircraft to the Chinese
Communists, threatening to cut off Marshall Aid and the Military
Assistance Programme to Britain unless its demands were met.
Under instructions from the British government, the Governor of
Hong Kong used administrative means to detain the aircraft for
several months. In the meantime, seven of the aircraft, which
were under police protection, were sabotaged by Kuomintang
agents. After some hesitation, Britain issued in May an Order-in-
Council which 'overrode the law as it stood and in effect made a
new law, which would inevitably pass the planes to the
Americans'.[87] Two years of litigation followed, after which the
aircraft were finally awarded to the United States.

   The Chinese Communist approach to the aircraft dispute was
distinctive because of their reliance upon the law of the colony
and their reluctance to confront the British authorities. At no
time and under no circumstances had they attempted to secure
the aircraft except by strictly legal means. The British fear that
the Chinese Communists might retaliate by fomenting strikes or
disturbances proved to be ill-founded. After the Chief Justice of
Hong Kong ruled in favour of the Communist government and
removed the injunction against the aircraft on 24 February 1950,

the Communist government could have ordered the planes to return to China. If the Governor had used further administrative measures to prevent the aircraft from flying, they could have been dismantled and removed by sea. The Communists could also have claimed the jurisdictional immunity enjoyed by a foreign state over the thirty-nine aircraft belonging to the Central Air Transport Corporation, which was a Chinese government department, but they chose not to do so. When the British government issued the Order-in-Council, the Chinese Communist legal representatives merely 'registered vigorous protest'; their attitude was 'one of sorrow rather than anger'. The Chinese Communist government, aware of its legal position, chose to stomach the injury and injustice because it did not wish to upset its relations with Britain.[88]

Although the Chinese Communists did not suffer an open defeat in the field of propaganda, as they did with regard to the Chinese state properties, they failed to gain new ground. Following the *Hwa shiang pao's* move to Canton in late 1949, the *Wen hui pao* became the Communist Party's organ in the colony, but this only contributed to its financial difficulties. The newspaper would have gone bankrupt in July 1950 had it not been bailed out by businessmen with Shanghai connections. The other leading Communist newspaper — the *Ta kung pao* — was, according to Eric Chou,[89] under the direct control of the Hsin Hua News Agency, which was 'some sort of control tower for the Communists in Hong Kong', but the paper 'never freed itself from bureaucracy and red tape, not to mention personal influence'. Worse still was the attitude of the senior members of the newspaper, most of whom 'talked like Marxists and yet acted and lived like capitalists'.[90] The popularity (or, rather, the limited popularity) at the time of the Communist newspapers is indicated by their circulation compared with that of other quality local Chinese-language newspapers. In mid-1950, the independent *Wah kiu yat pao* had a circulation of 46,000; the pro-Kuomintang *Kung sheung jih pao* circulated 22,000 copies (including 10,000 to Taiwan); the *Ta kung pao* circulated 12,000 copies; and the *Wen hui pao* circulated a mere 6,000 copies. The remaining Chinese-language newspapers, which were guided primarily by commercial motives, remained neutral or slightly anti-Communist. The government's expectation that these newspapers would gradually become pro-Communist in attitude thus proved to be ill-founded.[91]

As far as education was concerned, Communist infiltration was closely linked with the labour movement. The so-called workers' children schools — which the government described as 'hot-beds of Communist teaching' — were very much a product of the time. Chinese parents in the colony were extremely keen to send their children to school; education was the traditional means of achieving upward mobility. Some Hong Kong officials, very aware of this attitude, made commendable efforts to provide greater educational facilities for children, including those of the hitherto almost completely illiterate fishing community. However, the relative neglect of the pre-war years, the diversion of educational funds for the rebuilding of schools demolished during the war, and the rapid expansion of the local population since 1937 meant that, despite its efforts, the government still failed to provide adequate school places, particularly for children from working-class families. Even as late as 1950, only one-third of the 150,000 students in the colony attended schools which received any government financial support. Working-class parents had no choice but to send their children to schools sponsored by unions because they were unable to afford to send them to private schools, which were run mostly on commercial principles. A good opportunity was therefore afforded the local leftists to set up workers' children schools to spread pro-Chinese Communist teachings. The success of the leftists was nevertheless limited. The total enrolment in the workers' children schools in the mid-1950s is not known, but it has been estimated at between 10,000 and 12,000 (less than four per cent of the total student population of nearly 300,000, and less than forty per cent of the children of the more than 30,000 Federation members, who should in principle have furnished a considerably higher enrolment figure). On the whole, the local leftists failed to exploit the situation fully because the Chinese government did not intend to invest heavily in Hong Kong's young people and, in any case, the Hong Kong government prevented these schools from obtaining financial support from outside the colony. The Hong Kong government also curbed the Chinese Communist infiltration of education in two other ways: schools which could be proved to have provided Communist teachings were deregistered, and teachers and school administrators who were found to have Communist connections were deported. Such policies succeeded in keeping Communist infiltration at bay. In addition to the Tat Tak College, two workers' children schools were closed in 1949. Another heavy blow was the deportation in early 1950 of Lo Tung, the top Communist official on educational matters in Hong Kong and principal of the largest leftist school, the Hsiang Tao Middle School.[92]

In the field of intelligence, the Chinese Communists succeeded in building up a considerable network in Hong Kong. In the same way that the Western powers used the colony as a 'spyhole' on Communist China, the Chinese Communists used it as a 'listening post' for the outside world. As a result of the paucity of contact elsewhere, they depended heavily on intelligence gathered in Hong Kong to double-check information collected from such places as Geneva, Amsterdam, and Brussels. The importance of Hong Kong in this respect was confirmed in 1957 by Meng Chiu-chiang, head of the United Front Department for Tientsin and of the *Ta kung pao* organization in China. However, intelligence work of this kind was not a security risk to the colony; at any rate, it was no greater than that posed by the vast American intelligence network run by the United States' Consulate-General.[93]

The Chinese Communist policy with respect to the status of Hong Kong was similar to that of the Kuomintang, at least in the short term. According to the United States' Consul-General in Peking, Lo Lung-chi (Leader of the Chinese Democratic League, the most important organization of the so-called third force in Chinese politics), who had met Mao Tse-tung on 20 September 1949, was able to intimate indirectly that the position of Hong Kong was secure: the Chinese Communist Party had decided to accept as valid all treaties regarding Hong Kong's status which had been signed before the Kuomintang came to power.[94] Even the Foreign Office and Nicoll, the Officer Administering the Government in Hong Kong, concluded that the Chinese Communist policy for the colony was one of accepting the British position in Hong Kong, at least in the short term.[95]

In practice, the Chinese Communists were more conciliatory than the Kuomintang with regard to the British possession of Hong Kong. They had conspicuously avoided attacking the British authorities in Hong Kong before 1949. When they first criticized the Hong Kong government, in March 1949, they did so only by proxy, through Marshall Li Ch'i-shen (in Cantonese, Li Chai-sum) who was the leader of the Kuomintang Revolutionary Committee (another so-called third force organization). The attack was in response to the government's raid on the home of Lien Kuan, a senior Communist cadre, and the closing of the Tat Tak College. The second Communist propaganda attack, in June 1949, was again made by proxy, the issue being the introduction of the Societies Ordinance. The Chinese Communist propaganda machine only began to attack the Hong Kong government directly after Fang Fang's residence had been raided. The restraint of the Communists during this period was particularly remarkable because it coincided with the massive reinforcement of the Hong Kong garrison, which was seen as 'warlike' even by the British community in Shanghai.[96] The intensity of the Chinese Communist propaganda attacks against specific actions of the Hong Kong government subsequently increased, particularly in 1950, but the language used was, in Nicoll's terms, 'measured and unprovocative'. The Communists had consistently avoided challenging the British possession *of* the colony; they merely attacked the British administration *in* Hong Kong. This distinction was confirmed regularly by the Governor's half-yearly reports on Communist activities in the colony and by the observations of the Foreign Office.[97]

Despite an upsurge in the prestige of the Chinese Communists following the inauguration of the People's Republic and its recognition by Britain in January 1950, they made little headway in infiltrating Hong Kong, even if that had been their objective. On the whole, the Communists had been discreet in their activities and correct in their attitude towards the colonial government. There were, of course, a few exceptions. The sale of 'Victory Bonds' to Chinese businessmen in the colony had involved blackmail tactics of some kind, although this was as much an indication of their impotence as of their indiscretion. If they had had a strong influence amongst the local merchants, they would not have had to resort to such tactics.[98] Another notable example of 'Communist excesses' was a series of grenade-throwing incidents in May 1950 carried out by an ex-guerrilla unit of East River Column connection.[99] Once again, however, such acts betrayed the weakness, rather than the strength, of the Communists, having been committed in order to curry favour with the local Communists. Yet these acts, which could at best be described in the Communist jargon as 'leftist deviation of a venturesome nature', were against the policy of the local Communists. If the Communists had had effective control over the unit, the incidents would not have occurred. The question of appointment of a Chinese consul-general or successor to T. W. Kwok as Chinese Special Commissioner in the colony was an important indication of the attitude of the Communist government. Although it might have desisted from appointing a consul-general in order to avoid the public acceptance of British sovereignty over Hong Kong, this reaction would not have applied to the appointment of a special commissioner or representative (which would not imply Chinese acceptance of Hong Kong as a foreign territory).[100] The Chinese Communists probably simply had no intention of maintaining a public office which might be forced by local public opinion to confront the colonial government.

The rise to power of the Communist Party in China during this period was accompanied by a decline in the influence and prestige of the Kuomintang in Hong Kong (and, of course, in China). This decline was due largely to the loss of the support of a strong government across the border, but it was also due in part to the Hong Kong government which outlawed (amongst others) the Kuomintang by the Societies Ordinance in early 1949. After an unsuccessful appeal for special treatment, the Party wound up its local branch and ended the brief period of its legal existence in the colony following the end of the war. The decline of the Kuomintang in the colony indicated that it had ceased to be a formidable menace to security and order. Although the Colonial Office had wished to outlaw the Party as early as 1947, when Young was still Governor, it had been impractical to do so before 1949. The Kuomintang had also ceased to direct its activities against the government. Even its two main duties as the nationalist party of China — to protect Chinese national interests and to support Chinese nationalistic activities in Hong Kong — had been taken over by the Communists. Yet the threat to internal security posed by the Kuomintang was not a negligible one. Its activities against the Communists were as much a potential threat to the peace and order of the colony as the Communist activities against the Kuomintang government had been prior to October 1949.[101]

The greatest security problem which the Kuomintang continued to pose to the colony related to its covert activities against the Communists. In 1950, it had begun to form a guerrilla organization in Hong Kong under the direction of General Tsui Tung-loi, the guerrilla chief for east Kwangtung, for operations against the Communists across the border. In 1951, a series of raids was launched in the border region, but there is no evidence to prove that these operations were mounted in British territory. The success of this Kuomintang guerrilla organization in Hong Kong was limited, and it managed to exercise only doubtful influence over the guerrilla forces (which were usually also bandits) in Kwangtung. In any case, Tsui was assassinated, probably by his rivals in the Kuomintang, in March 1951 and the unit had to be reorganized. In June, following a police raid on the Kuomintang guerrilla headquarters, Major-General Wong Chung-man, the actual head of this force in the colony, and his entire staff were deported. Wong's successor, Cheung Yim-yuen, was also expelled from the colony in August 1951. More damaging to Hong Kong's security and order were the activities of General Kot Siu-wong, head of the guerrilla force for west Kwangtung. Kot attempted to lay an underground network against the Communists by forming a Triad society known as the '14 K' in Canton before its fall in October 1949. After the Communists occupied the city, the society migrated to Hong Kong and gradually became one of the most important Triads in the colony. Although the '14 K' challenge was primarily criminal rather than political in character, it provided a ready pool of ruthless supporters for the Kuomintang in large-scale violent confrontations with the Communists and, hence, was a potentially important security problem.[102]

Another potential threat to the law and order of the colony was the arrival from 1949 onwards of a large number of Chinese refugees. The fact that twenty per cent of these refugees had previous military training and some might still have arms was bad enough; the existence of a small, hard core of pro-Kuomintang refugees in the colony's refugee camp made matters worse. In fact, however, pro-Kuomintang refugees and pro-Communist workers clashed violently only once when, in June 1950, a group of workers sang pro-Communist songs during a picnic near the Mount Davis refugee camp on Hong Kong island. Subsequently, the government resettled the pro-Kuomintang refugees at Rennie's Mill camp, which was in an isolated spot north-west of the

Kowloon peninsula. Although no further incidents occurred, the Rennie's Mill camp became a stronghold of the Kuomintang, but it remained largely a potential rather than an immediate threat to local security. The Kuomintang activities in the camp were not reported in the Governor's half-yearly reports on Kuomintang activities — an indication that the Governor attached relatively little importance to the camp.[103]

The importance of the refugee problem in Hong Kong in the context of this study is that it was not treated as having any immediate relevance to the constitutional reforms then under consideration despite the size of the refugee population, and the large number of senior Kuomintang government, military, and party personnel included among them. With the exception of the Young Plan, which would have permitted such refugees to participate in municipal elections after they had been in Hong Kong for at least six years, no proposal considered by the British authorities made any provision for these refugees to participate in the administration of the colony, at any level. In any case, the Young Plan was dropped in 1949. Of particular importance is the fact that the refugee problem was fully understood by the policy makers concerned and no one referred to the situation when deliberating on the reforms.[104]

The only field in which the declining influence of the Kuomintang was not particularly noticeable was the press. Although its party newspaper — the National Times — had collapsed at the beginning of 1949, most of the local Chinese-language newspapers did not switch their support to the Communists; hence, the Kuomintang did not appear to be losing to the Communists in this field. This did not mean, however, that they were subject to Kuomintang influence. Since most of the newspapers were anti-Communist, they often appeared to support the Kuomintang, but an anti-Communist stance was, of course, not the same as a pro-Kuomintang stance. A more accurate measure of the support given the Party was the circulation of the pro-Kuomintang newspapers. The Hong Kong shih pao (or Hong Kong Times), which began publication in 1949 as a successor to the National Times, had a circulation of only 13,000, of which 12,000 were exported to Taiwan.[105]

In other fields, the decline of the Kuomintang was apparent. Prior to 1949, many local schools displayed Chiang Kai-shek's portrait and routinely saluted the Republic of China flag in the morning. After Chiang 'temporarily retired' from the Presidency of China in January 1949, his picture began to be removed. By the end of the year, not only had most of his pictures been removed, but many schools had also stopped saluting the Republic of China flag, which was very similar to the Kuomintang flag, indicating that they had ceased to support the Kuomintang government because it had lost the 'mandate of heaven' in China. The Kuomintang had no better luck in the labour field in 1949. In fact, it suffered a serious blow when the Hong Kong branch of the Chinese Seamen's Union, hitherto one of the most powerful unions in Hong Kong, broke away from the right-wing Hong Kong Trade Union Council. Thereafter, local unions tended to move away from the Trade Union Council without necessarily moving closer to the left-wing Federation of Trade Unions. Following the Federation's blunder in the Tramways dispute in 1950, the Trade Union Council was able to arrest its decline, but in general, the Kuomintang had lost much of its power in labour, as it had in education. By the end of 1950, its influence in both fields bore little relation to that exercised in 1946–7 at the height of its power.[106]

The closure of T. W. Kwok's office following Britain's official recognition of Communist China in January 1950 marked the end of an era during which the Kuomintang had held a special position in Hong Kong. The positions of the Kuomintang and the Chinese Communists in the colony had been reversed. Instead of being the ruling party — the government — in China, the Kuomintang had become a minority party openly opposed to the government of China. As far as the government of Hong Kong was concerned, the Kuomintang had also become less of a threat to Hong Kong's security, much in the same way that the Communists had posed little threat prior to October 1949.[107]

On the issue of constitutional reform, in 1949 both the pro-Kuomintang and the pro-Communist newspapers took a stand hardly distinguishable from that taken by other leading local Chinese-language newspapers when the Legislative Councillors put forward an alternative to the Young Plan. In general, they all recognized that there were differences of opinion as to the best means of reforming the constitution of the colony, and they considered the proposals put forward by the Chinese Reform

Association and 141 other representative bodies to be acceptable.[108] The only significant difference between the pro-Communist, the pro-Kuomintang, and the independent newspapers was the tone in which they addressed the issue. The two Communist newspapers (the *Ta kung pao* and the *Wen hui pao*) merely requested politely that the government respect the best interests of the majority of the people, that it avoid discrimination, and that it make good its earlier promise of introducing a municipal council without delay. The editorials of the pro-Kuomintang *Kung sheung jih pao* were harsher in tone, but only a little harsher than those of the independent *Wah kiu yat pao*. In any case, the target of the *Kung sheung jih pao* was not the government but the Unofficials, who had blatantly disregarded the expressed views of the overwhelming majority of articulate Chinese organizations. The editorials of these newspapers all discussed the reform proposals from a Hong Kong point of view. There is nothing in the editorials to suggest that either party was interested in securing an opportunity to infiltrate the proposed electoral machinery.[109] After the summer of 1949, it was the independent newspapers which reminded the government that the reforms had not been forgotten; both the pro-Kuomintang and the pro-Communist press remained generally silent on the reform issue, in which they were not particularly interested.

In the context of this study, of particular importance was not the fact that the government of Hong Kong faced a potentially very serious security problem (particularly if the Chinese Communists chose to challenge it), but the fact that the government was not apprehensive. Unlike the Colonial Office, which was concerned about the security of the colony in view of the rise to power of the Communists in China, Governor Grantham on the whole had assessed correctly the political situation and knew that Hong Kong was not in any real danger of being destabilized or attacked. It ought to be remembered that Grantham did not turn down the Young Plan for security reasons; he rejected it on 'practical grounds'. Indeed, Grantham was more interested in getting the Young Plan out of his way than in pushing through his own recommendations, which he believed to be safe. The potential challenge from either of the two leading Chinese political parties and the vicissitudes of local politics did not make him change his mind; Grantham simply did not think that major constitutional reform was appropriate for the colony.[110]

Economic Changes and Reform

Economic changes up until December 1950 had no important direct impact on political reforms in Hong Kong. Available statistics suggest that the economy of Hong Kong continued to grow at a breath-taking pace. The total trade of the colony in 1949 was HK$5,069 million (at current prices), or over thirty-eight per cent above the total value of trade in 1948. The figure for 1950 was HK$7,503 million. This represented a forty-eight per cent increase over the 1949 figure, or a rise of over sixty-six per cent above the 1948 figure. The number of registered industrial establishments increased in 1949 by 140 to 1,280 (or by almost thirteen per cent). It further increased by 242 to 1,522 (or by almost sixteen per cent) in 1950. The number of people employed in registered factories rose by almost seven per cent to reach 64,831 in 1949, and by almost thirty-eight per cent to reach 89,268 in 1950. The total consumption of electricity also increased by a yearly average of forty per cent in both years.[111]

This tremendous growth of the economy had no effect on constitutional reform for two reasons. First, economic development was not directly related to political reform, at least not in Hong Kong where the political system had remained basically unchanged for a century while the economy was transformed from that of a small fishing port to a modern metropolis. Secondly, officials in the colony and Whitehall fully recognized this fact and avoided linking the two issues. The most important factor which almost completely isolated political changes from economic development in Hong Kong was the trading post mentality. Both the British and Chinese traders regarded the colony merely as a trading post. They did not come to Hong Kong to settle down and raise their families; they came merely to make money. The overwhelming majority of British traders chose to return to England to live as country gentlemen after retirement. A large number of Chinese traders also preferred to retire to their home villages or towns in south China, although many did eventually settle in Hong Kong.[112] Until the permanence of the Chinese Communist regime in China was accepted in the early 1950s, many of the Chinese traders who had settled in Hong Kong still cherished the idea of returning to their home in China, especially upon retirement. Consequently, economic prosperity did not generally encourage either the

British or the Chinese traders to demand a greater share in local self-government.[113]

Given the uncertainty in the early part of 1949 over the future of the colony and the general political instability in the Far East during this period, an important question is whether the business community was so apprehensive about the situation that it encouraged the government to mark time with the proposed political reforms. While the business community certainly wished for a firm British commitment to maintain the status of Hong Kong and to safeguard its investments in the colony, it was on the whole cautiously optimistic, particularly after the local garrison was substantially reinforced in 1949. Neither the Kuomintang government's attempt to blockade Chinese ports under Communist control nor the outbreak of the Korean War had any significant immediate adverse effect on Hong Kong's trade; if anything, these events contributed to the revival of the China trade. The Communist Chinese government, which feared that it might be cut off from Western supplies, increased its purchase of all kinds of essential supplies through Hong Kong, particularly in 1950. As a result, the percentage of China trade in relation to Hong Kong's total trade increased from under twenty per cent in 1948 to over twenty-three per cent in 1949 and reached almost thirty-one per cent, its first post-war peak, in 1950. Despite the political situation, there is no evidence that the business community as such tried to influence the government either to expedite its handling of the reform proposals or to mark time.[114]

Despite the tremendous growth of the economy (including the number of people employed in registered factories) reflected in the statistics for the period, there was also significant unemployment, mostly amongst refugees and newcomers. The total number of unemployed and underemployed in the latter part of 1950 has been estimated at around half a million, including 200,000 who had previously served under the Kuomintang government in China. That is, one in four local residents was either unemployed or underemployed. Without social security of any kind, this situation might have caused social unrest and disorder had it not been for the Chinese family system and the fact that some refugees returned to China as the situation on the mainland stabilized. The unemployed who remained in the colony were generally taken care of as a moral duty by their relatives. According to the *Far Eastern Economic Review*, very few families in mid-1950 did not have one or more relatives staying as a guest while looking for a job.[115]

The situation in Hong Kong changed significantly, however, when the United States imposed strict restrictions on trade with Hong Kong in December 1950. As a result of Communist China's direct intervention in the Korean War, the United States imposed a trade embargo against Communist China. It also restricted trading with Hong Kong in order to prevent Communist China from using Hong Kong to evade the embargo. Although the full economic effect was not felt until after the end of 1950, these trade restrictions marked the beginning of the end of Hong Kong's sustained post-war boom in economic development. Of importance in the context of this study is the fact that the American trade restrictions foreshadowed an economic depression which was widely covered in the local press.[116] By February 1951, the situation had deteriorated to such an extent that the Governor was in a position to put aside all reform proposals while the attention of the colony was diverted to tackling the depression or expected depression. Economic changes had finally had an impact on the proposed constitutional reforms.

The Apathetic Speak Out

Prior to 1949, the various local organizations interested in reform were either waiting for the implementation of the Young Plan or were in favour of some undefined constitutional advancement. In 1949, many such organizations which had hitherto not expressed their views clearly or strongly began to do so, and there developed what might be called an outburst of opinion amongst them and the local newspapers at about the same time as the Legislative Council 'debated' the Young Plan. Prior to these developments, two political organizations (the Hong Kong Reform Club and the Hong Kong Chinese Reform Association) had been formed in January and May, indicating that the better-educated sector of the community had not lost its enthusiasm for reform despite the long delay since 1947.[117]

The Reform Club was formed by a group of professionals of various nationalities with the objectives of stimulating interest in public affairs and assisting the formation of a healthy public opinion; providing constructive general criticism of the government of Hong Kong; promoting better mutual understanding between the government, the legislature, and the people; 'foster[ing] and maintain[ing] an effective opposition' in the legislature; and devising ways and means of revitalizing and remodelling the Legislative Council 'with due regard to the present necessity for a Government majority'. The leading lights of the Reform Club were two British lawyers, Charles Loseby and Brook Bernacchi. Loseby was 'horrified at the political set-up and doubly horrified' by the way in which the government had handled the question of reform. Loseby and his friends saw the crux of the matter as being the Young Plan's proposal to split the population into separate Chinese and non-Chinese electorates, which could create racial tension. They also thought that the introduction of the electoral principle into the Legislative Council was the best approach to reform in Hong Kong.[118] The Reform Club was potentially the nucleus of a political party with generally liberal political views.

In early 1949, the Reform Club took the initiative in discussing its counter-proposals to the Young Plan with the unofficial members of the Legislative Council. On the day of the Legislative Council 'debate', it submitted its own proposals to the Governor in the form of a petition. Briefly, the Reform Club called for a two-stage reform of the Legislative Council and the re-examination of the municipal council scheme. The Club proposed that an interim Legislative Council of twenty members be set up, with eleven unofficial and nine official members, for a period of between twelve and eighteen months. Of the eleven unofficial members, three would be nominated by the Governor, while the remaining eight would be directly elected by a joint electorate of British subjects and Hong Kong citizens. (A Hong Kong citizen was defined as anyone who had resided in the colony for at least five years, and who had signed a written undertaking to uphold the interests of the colony as being of paramount importance and to defend the colony as might be required and directed by the government.) The Club also proposed that the interim Legislative Council proceed to examine the municipal council proposals with a view to putting forward 'an improved Municipal Council Scheme'. At the end of the interim period, which was required for the complete registration of Hong Kong citizens, a new permanent Legislative Council would be constituted with nine official and eleven unofficial members, all of whom should be directly elected.[119]

The proposals of the Reform Club were in several respects similar to the suggestions embodied in Landale's motion of April 1949: both the size of the proposed Legislative Council and the size of the Unofficial majority were the same. According to Bernacchi, however, the Reform Club did not get its idea of reform from the Unofficials. The Reform Club had already drafted its own proposals before it discussed the matter with the Unofficials, who do not appear to have made any firm suggestions at that time.[120]

Notwithstanding the similarities, the Reform Club's proposals differed from those of the Unofficials in two important respects. First, the Unofficials' approach restricted the vote to British subjects, while the Reform Club's proposals included the proposed Hong Kong citizens — a category which could easily have expanded within a few years to include a large number of local residents. Although the Club's suggestion (unlike the Unofficials' scheme) represented a departure from the usual practice in British colonies at that time, it was in line with the promise made in 1946 by the British government to give the local inhabitants, not just British subjects (who formed at that time a small minority), greater local self-government. Secondly, the Unofficials suggested that the nomination system for choosing unofficial members to the Council be retained, whereas the Reform Club called for direct election of all Unofficials to the permanent Legislative Council. The Unofficials and the Reform Club also differed on the size of the Council and of the Unofficial majority, but this difference was due to the Unofficials' amendment of their earlier proposals just before they 'debated' the Landale motion. Editorials, news reports, and letters published in the local newspapers suggest that the Reform Club's proposals were relatively more popular than the alternative proposed by the Unofficials.

The Chinese Reform Association, which had a predominantly Chinese composition, was formed by professionals and business-men in order to participate in the discussion about constitutional reform. According to its founding Chairman, Dr Wong San-yin (previously a lecturer in Pharmacology at the University of Hong Kong), the Chinese Reform Association did not intend that its members should run for elected posts when direct elections were introduced. The Association was not formed as a political counterweight to the Reform Club. It was founded mainly by Ma

Man-fai, who had felt insulted by his European colleagues when they formed the Reform Club. Whatever reason Ma might have had for setting up the Chinese Reform Association, it was not political differences with his colleagues in the Reform Club. In the first two or three years of its existence, the Association was not under Chinese Communist control, even though a few of its leading members (notably Mok Ying-kwai and Percy Chen) were connected with the Chinese Communists. It did not pass firmly under Communist control until 1952 or 1953.[121] During this period, the Association was also not, as the Colonial Office mistakenly believed, a 'one-man show'.[122] There were at least two factions in its Executive Committee, only one of which was pro-Communist. Secondly, the pro-Communist faction does not appear to have gained the upper hand in the struggle for the leadership of the Association until the end of 1951. On the whole, according to the Chinese-language newspapers, the Club received reasonably strong support from most of the Chinese organizations interested in reform in 1949.

Before the formation of the Chinese Reform Association, local Chinese who were interested in reform looked for the most part to the Hong Kong Chinese General Chamber of Commerce to speak for the Chinese community outside the Legislative Council. There was, however, a division of opinion on the question of reform in the Executive Committee of the Chamber, which in many ways reflected the division of views within the local Chinese community. On the one hand, Ko Cheuk-hung, Chairman of the Hong Kong Chinese General Chamber of Commerce (who had recently been awarded an honorary MBE), generally supported the view of the Unofficials following his consultation with Sir M. K. Lo (who also happened to be Legal Adviser to the Chamber). He further stressed that reform of the Legislative Council would be an acceptable substitute for the introduction of a municipal council.[123] On the other hand, some members of the Executive Committee were reluctant to embrace the Unofficials' ideas as the Chamber's own. They were sceptical of the idea of abandoning the municipal council and wanted representation of the non-British Chinese taxpayers in the Legislative Council and the establishment of an elected municipal council. Ko eventually managed to have his views adopted by the Chamber, although they did not receive widespread support from those Chinese organizations interested in reform. As a result, the Chinese

Reform Association was able to assume leadership of the articulate Chinese in this matter.[124]

In mid-June, Sir M. K. Lo's intention to amend Landale's motion became known to the press, provoking considerable criticism from the Chinese Reform Association and various local chambers of commerce, with the exception of the Hong Kong Chinese General Chamber of Commerce. They strongly condemned Lo's proposal to abandon Young's municipal scheme. Indeed, it was the passing of Lo's motion in the Legislative Council on 22 June which, more than anything else, had stimulated the outburst of opinion by these organizations. According to the *Kung sheng jih pao*, there was a strong feeling among those interested that the Unofficials had disregarded both the views and interests of the Chinese community. On 30 June 1949, in an open meeting convened by the Hong Kong Chinese General Chamber of Commerce to discuss political reform, the Chamber was the only organization which supported in principle the Unofficials' alternative. All of the participants who represented organizations other than the Chamber, and who spoke at the meeting, disagreed with the views of the Chamber and criticized the way in which Ko had handled the matter. This division of opinion between the Hong Kong Chinese General Chamber of Commerce and most of the other articulate Chinese bodies was used extensively by the Unofficials and the Governor to dismiss the views of the other Chinese organizations which were critical of the Unofficials' alternative. Following the 30 June meeting, the Chinese Reform Association, which had been very outspoken on reform, collaborated with the Chinese Manufacturers' Union, the Kowloon Chinese Chamber of Commerce, and the Kowloon General Chamber of Commerce in inviting the leading representative bodies of the Chinese community to discuss further the question of constitutional reform. Although the Hong Kong Chinese General Chamber of Commerce was invited, it declined the invitation.[125]

On 13 July, 138 representative bodies of the Chinese community joined the Chinese Reform Association and the three other organizers in a public discussion on political reform. The meeting approved unanimously a widely publicized draft petition on reform. Although all eight points in the petition were proposed and seconded by different bodies, they were based on the views expressed earlier in the press by the organizers. The petition dealt with both Legislative Council reform and municipal reform. On the whole, the views of these organizations were not very different from those of the Reform Club; their views on the Legislative Council reform were in fact very similar. They differed mainly on two counts. First, the Chinese organizations insisted that the Chinese community should have a majority of the Unofficial seats in the Legislative Council. They also demanded a fully elected municipal council, about which the Reform Club did not feel strongly. The Chinese petitioners also specifically called for the immediate expansion of the Legislative Council as an interim measure for one year; the registration of voters within one year; the introduction of a council of twenty members, to be constituted by nine official and eleven directly elected unofficial members of whom six should be of the Chinese race (regardless of national status); and the introduction of an Unofficial majority of not less than two. With regard to the municipal council, they requested that the government fulfil its promise to introduce a municipal council before May 1950; that it amend the draft Municipal Council Ordinance so that all thirty seats in the proposed council would be filled by unofficial members directly elected within the boundary of the proposed municipality as one constituency; that it avoid discrimination on racial or nationality grounds in all electoral matters; and that it appoint immediately a commission of eleven members (of whom six should be Chinese and five should be non-Chinese) to examine questions relating to the transfer of authority to the new municipal council with a view to promoting local self-government. The petition, which received the editorial support of almost all of the leading local Chinese-language newspapers, was submitted to the Governor on 16 July. The Governor promised a reply within about three months.[126]

Despite the fact that many of the participating organizations were chambers of commerce and manufacturers' unions, their main considerations were not economic in nature. They had made it clear that they did not support the Unofficials' alternative as such; rather, they had become outspoken because of their strong opposition to it. In expressing their views on reform they had also demonstrated that they were not irresponsible or unreasonable, or incapable of understanding the issues involved. Like the Reform Club, these organizations had asked for a departure from the usual British practice in reforming the

colonial legislature, but one which was in line with the expressed
policy of the British government. It is noteworthy that they
proposed, as a condition to the introduction of direct election of
all Unofficials to the Legislative Council, that the Unofficial
majority be limited to two — a provision which ensured that the
government could be defeated if and only if all the Unofficials of
whatever nationality presented themselves at one time and
voted unanimously against the government. The officials of the
Colonial Office found this condition to be unacceptable since it
implied that the government could be defeated.[127]

The demands by these organizations for the elimination of
racial discrimination, for communal representation, and for
indirect representation in the proposed municipal council were
entirely justified in principle; they amounted merely to a demand
for fair play. In practice, however, they meant that most of the
built-in safeguards which Young and the Colonial Office had
deemed essential if a municipal council were to be introduced
into the colony would have to be set aside; hence, they were
totally unacceptable to the Colonial Office.

After their views had been made clear, the articulate bodies
then waited for both the Hong Kong and British governments to
respond. However, for reasons explained earlier in this chapter,
the wait turned out to be a very long one and they were still
anxiously awaiting a positive reply in the summer of 1950. As
mentioned earlier, both the Reform Club and the Chinese
Reform Association enquired about the progress of the reforms
in 1950, and occasional editorials in the local newspapers also
indicated that further steps were expected to be taken about the
reform. The Reform Club went so far as to work out a comprehensive political platform and to prepare itself to participate
as a political body in the expected municipal council elections of
1950.[128] All of the available information suggests that the leaders
of these organizations were unaware of the turn of events in
Whitehall; they were waiting under the assumption that the
government was making all of the necessary preparations for
reform.

Whilst the majority of the residents of Hong Kong had
expressed no opinion on the various proposals for reform during
this period, the co-operation of well over one hundred local
organizations in the drafting, discussion, and submission of a
petition for reform — which would not benefit them directly —

was unprecedented in the history of British Hong Kong. Up until the early 1950s, with the exception of the strikes of the 1920s (which were related either to Chinese politics or to events in China), at no time had such a number of local Chinese organizations taken the initiative in associating with one another publicly on a local issue. Neither the controversy in 1946–7 over the introduction of income tax nor the Kowloon Walled City Incident of 1947–8 had resulted in similar actions. The outburst of articulate opinion in the summer of 1949 marked the awakening of civic consciousness among the local community leaders. If substantial progress had been made in the direction of introducing either the Young Plan or the Unofficials' alternative soon after the summer of 1949, it is possible that a high level of interest amongst the local organizations might have been maintained and political awareness might have spread.

6 Anticlimax

ON 20 November 1951, after the question of constitutional reform had been put aside for almost a year, the Governor and the Executive Council re-examined the issue in connection with the scheduled visit of the new Conservative Secretary of State, Oliver Lyttelton. Although, according to J. F. Nicoll (Colonial Secretary at that time), the 'responsible and professional' people (that is, the Unofficials) in the colony had not demanded reform, 'a vocal minority' at a 'lower level' who 'could not be classed as irresponsible' (presumably referring to the Reform Club and its supporters) were strongly demanding reform. Constitutional changes in other British colonies, such as in Singapore and the Gold Coast, had stirred up public interest. The general feeling in the colony was that while radical reform was out of the question, there should be some changes in the Legislative Council, which represented only a section of the population. Both Grantham and Nicoll recognized that having promised reform since 1946, the government could not delay it any longer. They also recognized that restiveness in Hong Kong itself would be more embarrassing than any Chinese Communist propaganda which might be provoked by the publication of the reform proposals that Grantham and the Colonial Office had agreed in the winter of 1950. The forthcoming first-ever visit by a Secretary of State for the Colonies was expected to be an occasion for just such an outburst of opinion among the interested local organizations.[1] Consequently, the Executive Council, which felt that the government was committed to a measure of constitutional reform only under 'normal conditions', agreed to examine the issue again even though 'conditions could not yet be described as normal'. It is difficult to determine precisely what the Council meant by 'normal conditions' because it had never defined the term; nor had it explained in what ways the situation was abnormal. According to the Executive Council minutes of the second half of 1951, the only issue which received much attention was the squatter or housing problem, which although very serious was not a new problem in post-war Hong Kong. The Unofficials might have been referring to the economic situation which was exceptional in that Hong Kong was facing its first economic depression in the post-war era. However, this seems unlikely in view of the Unofficials' lack of enthusiasm for reform even when the economy was booming. Most of the Unofficials simply did not want any constitutional reform. In fact, the Executive Council meeting of 20 November was devoted to examining how to advise the Secretary of State to dodge questions about reform which were likely to be raised by local community leaders and journalists during his visit. The Council did not discuss any of the reform proposals, nor did it invite W. J. Carrie, the Special Adviser responsible for constitutional reform since November 1948, to attend the meeting.[2]

On 12 December 1951, Lyttelton arrived in Hong Kong for his three-day visit, during which he met the Unofficials and four local delegations. Before his meeting with the local delegations, he was told by the Unofficials that the two reform clubs did 'not represent any genuine strength of public feeling and quarrel[led] too much amongst themselves to be regarded as responsible bodies'.[3] The way in which the Unofficials explained this comment is unclear because the original record of the meeting is still closed. They may have been referring to the division of opinion within the Chinese Reform Association or, mistakenly, to quarrels between the Chinese Reform Association and the Reform Club. Despite their differences as to the best means of reforming the government, there is no evidence that these two organizations engaged in a public quarrel over the reform issue. In any case, the Unofficials' view contradicted Nicoll's observation that the vocal minority calling for reform could not be classified as irresponsible. T. N. Chau, the senior Chinese Unofficial, also added that the Chinese community wanted only peace and security, not constitutional advance. While this was presumably true of the vast majority of the population who remained silent, it was not very relevant since the reforms then under discussion clearly excluded the silent majority (who were mostly non-British Chinese subjects) from having any role to play. Chau, his Unofficial colleagues, and the Governor apparently did not mention to Lyttelton the support which the two reform clubs received on the question of reform in the Chinese-language press. Lyttelton, who had hitherto not given much thought to the issue, generally found it 'sensible' to develop the Hong Kong constitution gradually, but he agreed 'to go no further ... than the minimum which might be considered necessary'. Unlike his Labour predecessors, Lyttelton was very willing to follow whatever advice his officials (and the unofficial members of the Executive Council) had to offer on the issue.[4] Of the four local organizations which met Lyttelton, only the delegation representing the expatriate-British-dominated Hong Kong General Chamber of Commerce did not ask for reform; the Chinese Reform Association, the Reform Club, and the Hong Kong Chinese General Chamber of Commerce all did so. According to the reformers, the fundamental political problem in Hong Kong was the complete separation of the government and the people. The nominated Unofficials, who were generally seen as 'yes men', had failed to bridge the gap between the government and the governed. The solution they suggested was the introduction of elected Unofficials into various local councils. On this occasion, in sharp contrast to the division of opinion among the Chinese organizations led by the Chinese Reform Association and the Hong Kong Chinese General Chamber of Commerce in 1949, both bodies changed their previous positions and asked separately for the implementation of the Young Plan. They both argued that this scheme would receive the widest support from the local Chinese and that it would fulfil the government's promise of reform made in 1946. This change in position was certainly not due, at least so far as the Chamber was concerned, to any change in its leadership. Still under the chairmanship of Ko Cheuk-hung, the Chamber no longer supported the Unofficials' alternative; it now lauded the Young Plan as a scheme which showed insight. The reason why the Chamber changed its position remains unclear, but it might simply have been because on this occasion (unlike in 1949) Ko had not consulted Sir Man-kam Lo before expressing his views. The Chamber also requested that the government let it nominate one of its members to sit on the Legislative Council. This was not an entirely novel idea since the European-dominated Hong Kong General Chamber of Commerce had been a nominating body for the Legislative Council since 1884. The Chinese Reform Association must also have recognized that its proposals of 1949 had no chance of being accepted. Hence, it supported the Young Plan which had already been approved by the British government in principle in 1947. It might also have felt that the majority of the local people would support the Young Plan in the changed circumstances, since it went so far as to invite Lyttelton to hold a plebiscite on this question in order to ascertain whether it was in fact correctly representing the view of the majority of the people.[5] The primary concern of the Reform Club was to reassure Lyttelton that it was responsible, that it represented the interests of the various nationalities in the colony, that it understood the delicacy of the existing political situation, and that its proposals set out in its 1949 petition were safe.[6]

Of those newspapers surveyed, the English-language *South China Morning Post* alone did not favour the implementation of any of the reform schemes publicly discussed since 1946. It conceded that taxpayers should in principle be properly represented, but it argued that reform should be introduced only if it would bring 'more effective provision and administration'. Specifically, it held that constitutional reform was less immediate a problem than were security, livelihood, housing, health, and education. Whilst it might have advocated that the government pay more attention to such material problems, its argument inevitably led to the conclusion that Hong Kong should not have any constitutional advancement for some time to come. Since the colonial government had up until that time refused to provide public housing on any scale, it would be decades at least before that problem could be tackled adequately. However, in view of the fact that the lease for the New Territories would end in 1997, the postponement of reform for any length of time would mean in effect that it would be rejected. This was a minority view, and the editor of the newspaper published a second editorial solely to defend the first. This second editorial argued that the government of Hong Kong had been very efficient and that it would be unwise to replace government by professionals with government by amateurs. The editor appears to have missed the point, however. There was no question of elected (or nominated) deputies replacing senior civil servants as the heads of administrative departments. The crux of the constitutional reform issue in Hong Kong at that time was the means by which the local inhabitants, who were mostly non-British, could be given a say in the management of their own affairs without affecting the status of Hong Kong as a British colony.[7]

The editorials of the leading independent Chinese-language newspapers indicated that they favoured some form of reform, but that they also recognized that the long delay would mean they would have to accept something less than they had asked for two years before. In contrast, neither the Kuomintang-dominated nor the Communist-controlled newspapers referred to reform at all; they were presumably no longer interested in the matter.[8]

The newspapers and public organizations which expressed any views had all been loudly in support of reform in 1949, and the cry for reform was on the whole relatively weaker. This was partly because the issue involved at this stage was non-controversial, therefore evoking less discussion and attention. However, it was also partly the result of the depression. The attention of most of the original 142 Chinese organizations which had demanded reform in 1949 had been diverted to the more pressing and immediate question of the survival of their member firms. Nevertheless, in addition to such public bodies, a few individuals had also written to the local newspapers to 'demand the immediate grant of constitutional reform'. It is, of course, doubtful whether such individuals could have represented the views of the people of Hong Kong, yet the fact of their writing to the newspapers suggests that the idea of reform did receive some support from people other than the members of the articulate public bodies.[9] However, the extent of the support for reform among the silent majority cannot be ascertained.

Although, during his visit, Lyttelton publicly assured the colony that the reforms were receiving his 'sympathetic consideration', he dodged all specific questions by saying that he was there to study the problem at first hand. Privately, Lyttelton and J. J. Paskin (Assistant Under-Secretary supervising Hong Kong affairs) reached an understanding with Grantham on what to do about reform. Since Grantham had reassured Lyttelton and Paskin that the proposals agreed in the winter of 1950 were still 'safe', they agreed to take those proposals out of cold storage. It was also understood that the Governor would begin by sending a semi-official letter to Paskin. This was done in January 1952 after the Executive Council had been duly consulted.[10]

J. B. Sidebotham, Head of the Hong Kong Department, and his assistant in the Colonial Office strongly opposed the implementation of the reform proposals agreed at the end of 1950. They preferred to let sleeping dogs lie and, if the Americans were to blockade Communist China as a result of the extended conflict in Korea, to 'put to bed for good' the question of constitutional reform in Hong Kong. They argued against reform on two grounds. First, the people of the colony, particularly the Unofficials, did not want any major change. Furthermore, since the Governor had agreed to put aside the reforms in February 1951, the Far Eastern situation had deteriorated considerably and local morale had been further shaken by virulent propaganda attacks from China.[11]

While Sidebotham and his assistant thought that they had good reasons for raising their objections, their reasoning was based on observations which were not entirely correct. According to the monthly political intelligence reports, public morale fell to its lowest ebb at the beginning, rather than at the end, of 1951 when the United States precipitately advised its citizens to leave the colony. Moreover, by January 1952, the *Far Eastern Economic Review* felt confident enough to state categorically that 'Hong Kong's nervousness [had] gone, no longer [was] there any talk of Chinese Communist aggression aimed against the Colony'. It added that even Hong Kong's commercial future was not a problem. Sidebotham and his assistant were also surprisingly ill-informed as to the articulate opinion on reform during Lyttelton's visit. For instance, the Colonial Office had no record of the fact that the Chinese Reform Association had pressed for reform when its deputation met Lyttelton, even though the Association discussed only political reform with the Secretary of State. This issue was widely reported in all the leading Hong Kong newspapers and it was not contradicted by the Hong Kong government. Sidebotham was also incorrect in thinking that both the Reform Club and the Hong Kong Chinese General Chamber of Commerce (which, on the information available to him, were the only two organizations asking for reforms) were 'closely linked'. The two organizations were not generally closely connected and their views and demands on constitutional reform differed. Sidebotham's arguments against reform were therefore ill-founded.[12]

The queries of Sidebotham and his colleagues were answered fully in February by J. F. Nicoll, who was on leave in England prior to his taking up the governorship of Singapore. Nicoll made it absolutely clear that the objections to reform raised by the Foreign Office in early 1951 (namely, that the introduction of the 'undemocratic' reforms then under consideration would only provoke the Communist authorities in China to launch a propaganda attack against Hong Kong) were no longer valid. The Chinese Communist propaganda machine had already intensified its attack against the colony since Winston Churchill had become Prime Minister in October 1951. Nicoll also stated categorically that 'there was nothing to worry about' in relation to the size of the proposed Unofficial majority, and that the government could carry the day on any important issue. In fact, the existing Finance Committee of the Legislative Council already had an Unofficial majority and the government had never been defeated. Instead of weakening the internal security of the colony, Nicoll was confident that the introduction of these changes would help the government because they would satisfy the vocal supporters of reform.[13] He had left no grounds on which the officials in the Colonial Office could oppose the reforms.

Lyttelton and Paskin, who discussed the issue with Grantham in Hong Kong, did not share Sidebotham's conclusions. They agreed with Grantham and Nicoll that the introduction of the 1950 reform proposals would be a lesser evil than political agitation for reform in the colony. However, since Sidebotham had warned that the proposal to change the Legislative Council was a major reform, and had indicated that he would proceed with the matter only if expressly instructed to do so, they opted for caution. Before making a decision, Lyttelton put the matter to the Prime Minister. He was, like Paskin, indifferent to the Hong Kong reforms.[14]

When the Foreign Office, the Commonwealth Relations Office, the Ministry of Defence, and the Treasury were consulted, the Foreign Office alone offered serious comment. The Foreign Office continued to take the view that these 'undemocratic' reforms, which were 'by no means far-reaching', would provide ample justifiable grounds for a Chinese Communist propaganda attack. However, it also accepted Grantham's and Nicoll's point that its objection to the reforms in early 1951 was no longer valid and that further delay in introducing these 'quite harmless' reforms would cast doubt on the good faith of the government. On balance, the Foreign Office agreed to raise no objection to the introduction of the reforms on the understanding that the Colonial Office could put the matter into cold storage again if at a later date the political situation in the Far East made it necessary.[15]

On 20 May 1952, the British Cabinet discussed the Hong Kong reforms. Lyttelton laid heavy emphasis on the fact that the proposals to introduce indirect elections and an Unofficial majority in the Legislative Council were safe. The Cabinet was interested only in the complete lack of reference to the interests of the Indian community in Hong Kong. While it recognized that the Indian government wished to protect the interests of the several thousand-strong Indian community in the colony, it was unwilling to give the Indian community special representation in the Legislative Council. It insisted merely that the Indian government and other interested Commonwealth governments be kept in touch with the progress of the reforms. The Hong Kong reforms were thus approved without any controversy.[16]

In the meantime, on 25 March, Grantham and the Executive Council decided to reintroduce direct elections into the Urban Council, which had been in abeyance since Hong Kong fell under Japanese occupation in 1941. They also decided that this information should be made known to the public by means of a question and answer in the Legislative Council. Both the question (which was later asked by T. N. Chau, the senior Chinese Unofficial) and the answer were drafted by the Attorney-General. They were duly made public in the Legislative Council on 9 April.[17] On 30 May, the Urban Council elections were held. Less than three months had been needed to prepare for the elections after the Executive Council decided in favour of their reintroduction, indicating that the government had the means of overcoming within a period of only a few months the practical problems involved in introducing some form of direct election for an electorate of over 10,000 people.

The Urban Council elections were the only public elections held in Hong Kong during the period under study. Despite the short preparation period, during which voters had to be registered and candidates nominated, there were nine candidates for the two vacancies; all were men of high calibre. One candidate, G. S. Kennedy-Skipton, was a Class One Cadet and Chairman of the Urban Council before the war. The others were all successful professionals or businessmen. With the exception of Kennedy-Skipton, who as a 'neutral' Irishman had continued to serve in Hong Kong during the Japanese occupation, and Daniel Chen, who was educated in China and the United States, all of the candidates had either received their education and/or their professional training in Britain or Australia, or had served the British during the war.[18] With the possible exception of the pro-Communist candidate, Percy Chen (a successful English-trained barrister), their backgrounds would not appear to indicate that they preferred that Hong Kong cease to be a British colony and become part of Communist China or of Kuomintang-controlled Taiwan. In fact, despite Chen's close connection with the Chinese Communists, he was emphatic that he would confine himself to local affairs. He campaigned against 'yes men', corruption, and favouritism, and supported open discussion of civic affairs. Although Chen was at the time Vice-Chairman of the Chinese Reform Association, he campaigned as an individual, rather than on the Association's ticket, and was believed to be one of the frontrunners. The Hong Kong Reform Club, for its part, showed considerable political sophistication in its campaign for its two candidates, Brook Bernacchi and P. C. Woo, who campaigned together on the Club's ticket, putting up a common platform which they vowed to carry out if elected. Their main concerns were the development of the Urban Council into a council with real responsibilities and the improvement of the general living standard of 'Mr and Mrs Hong Kong'.[19]

Of the 9,704 registered electors, only 3,368 (just under thirty-five per cent) voted at the elections. The low turn-out was due partly to the fact that there was only one polling station for the entire colony. One of the Returning Officers admitted that the lack of a polling station across the harbour in Kowloon had contributed to the small number of votes. Many of the 7,800 jurors who were automatically enfranchised would have been unable or reluctant to spend hours travelling from their place of work or residence in Kowloon, or in a few cases, the New Territories, in order to cross the harbour to cast their votes. It is also doubtful whether all of the qualified jurors were aware of their right to vote. The government's decision not to set up an additional polling station despite the criticism that appeared in the newspapers suggests that it probably did not want to encourage public participation.[20] The electoral list — comprising the juror list and other people who were qualified but who for various reasons were exempted from jury service — was also unsatisfactory. Whilst it was proper to enfranchise the several thousand jurors, they were hardly representative of the two million inhabitants of the colony. More importantly, it meant that the franchise was restricted to people literate in the English language, this being a prerequisite for jury service. In effect, such an arrangement excluded such people as wealthy Chinese

businessmen who had all or most of their investments in the
colony (and who might even be British subjects) but who did not
know the English language.[21] Another reason for the small number of votes cast was the fact that the Urban Council was in effect
merely an advisory board on sanitary matters; the elections were
not to positions of real power. It is possible that had the elections
been to the Legislative Council, the turn-out might have been
considerably higher. However, on balance, the elections were not
an unsatisfactory start, particularly since they were carried out
during a time of economic hardship.[22]

The really remarkable thing about the elections was the lively,
responsible, and reasonably sophisticated electioneering, which
suggests that those who participated in the elections were capable
of taking part in fair public elections. Prior to election day, even
the hitherto sceptical *South China Morning Post* had conceded
that the electioneering and the public reaction to it had placed
constitutional reform in 'a perspective appropriate to Hong
Kong's conditions'. On polling day, it went further, predicting
that the day would be remembered as 'the beginning of a new era
— for the competition conveys not only an unprecedented public
interest but a public determination to reform the system, increase
the public representation, and broaden the procedure of municipal administration'.[23]

When the results of the election were known, Chen had failed
to be elected. Despite his campaign rhetoric, Chen was the candidate closest to the Chinese Communists and his Communist connection had proved to be a liability rather than an asset.
Although he had been expected to be elected, he polled less than
500 votes, while the first three candidates each polled over 1,000
votes. Those candidates who campaigned individually also failed
to gain the support of most voters. The successful candidates,
Bernacchi and William S. T. Louey, campaigned on the tickets of
the Reform Club and the Kowloon Residents' Association respectively. It is worth noting that the Reform Club, which campaigned as a liberal party for Hong Kong, received relatively
strong support at the poll. Its second candidate, Woo, polled
third, losing the election to Louey by only thirty-seven votes.
This meant that the electorate well understood the benefit of
voting for a quasi-political party which vowed to carry out its
campaign promise, rather than for individuals who might be
unable to carry out their pledges. Of more significance was the

fact that the elections were non-racial: Bernacchi was originally an expatriate Briton and Louey was an Australian-born Chinese.[24] The fear of British officials both in Hong Kong and in the Colonial Office that elections in Hong Kong would be run strictly on racial lines was thus proved to be ill-founded.

Although it had never been intended, the Urban Council elections turned out to be a turning-point in the attempt to introduce constitutional reform in post-war Hong Kong. The small number of votes cast, and the fact that the elections were not widely discussed by a majority of the leading Chinese-language newspapers, led the Unofficials and the Governor to reconsider the whole question of reform, which they had been proceeding with merely as a price to pay in order to satisfy the articulate supporters of reform.[25] The Urban Council elections had provided grounds to believe that the price could be lowered. The Unofficials, who were almost all successful businessmen, found the opportunity too good to lose. Consequently, they attempted, in Sidebotham's words, to 'put to bed for good' the Legislative Council reform.

The Unofficials' volte-face occurred in mid-June, shortly after the Hong Kong government had finished analysing the Urban Council elections and had learned of the British government's approval of the reforms. On 17 June, the Executive Council decided to ask the Colonial Office not to proceed with the publication of dispatches between Hong Kong and London on the matter, at least until after the Governor had returned from leave in England in October. No reason for this decision was given in the minute of the meeting. The Governor later explained to the Colonial Office that, with the notable exception of the Portuguese Councillor (Leo D'Almada E Castro), the Unofficials had never been 'wholeheartedly in favour of reform' and that after the Urban Council elections they had become strongly opposed to any major change. The voice of the supporters of reform had been significantly weakened. On 24 June, the Executive Council met again and decided that 'any far-reaching changes in the constitutional set-up in Hong Kong at the present time were inadvisable and that changes should be limited to increasing the number of elected members on the Urban Council from two to four (with a corresponding reduction in the number of ex-officio members), and to giving the Urban Council greater financial autonomy within the limits of a block grant'. At the Unofficials' request, the Governor also agreed to persuade the British government to abandon the Legislative Council reform when he went on leave in England in July. As an initial step, Grantham sent a telegram to Sidebotham towards the end of June warning him that the Executive Council had become apprehensive about the Legislative Council reform, which would constitute a 'danger'. He outlined the Executive Council's proposals to limit changes to the Urban Council and suggested that the matter be discussed after he went on leave.[26]

The minutes of both Executive Council meetings in June made no reference whatsoever to the 'danger' of the reforms. Indeed, the absence of the Special Adviser for Constitutional Reform, the Commissioner of Police, and the Political Adviser (the Hong Kong equivalent of a Secretary for External Affairs) from the Executive Council's discussion of the matter suggests that neither political problems and internal security within the colony nor relations with China were involved. The 'danger' to which Grantham referred in his telegram to Sidebotham could not therefore have been a danger to the colony itself. When Grantham discussed the matter with the Colonial Office in July, he insisted that the proposed Unofficial majority in the Legislative Council was still 'safe' and that 'the Government could count on a majority vote on all major questions'. He explained that the word 'danger' was used to mean that certain elected Chinese members, if they were to be allowed in the Legislative Council, might speak in favour of the Communists or engage in discussions aimed at causing 'embarrassment and even disaffection', and might 'place other Chinese members in a difficult position'.[27] If the Legislative Council were to be used as a platform for pro-Communist propaganda, this would also be an embarrassment to the government. Although neither Grantham nor members of the Colonial Office had referred to this point in the accessible documents, it is logical to assume that they were aware of it. In other words, the 'danger' to which Grantham referred was primarily the potential risk of embarrassment to some of the Chinese unofficial members in the Legislative Council, and presumably also to the government.

While such a risk could not be ruled out, it remains difficult to explain why this should have been a major concern in 1952. This element of risk had been inherent in the situation since the Communists came to power in China in 1949. According to

Grantham, even the 1949 proposals for a large Unofficial majority by direct elections were safe as long as the electorate and
membership of the Council were limited to British subjects.
Under the 1950 proposals — which were those being discussed in
1952 — all unofficial members still had to be British subjects and
they were to be indirectly elected by the rich and educated. The
only difference between the 1949 and 1950 proposals was that
two of the electoral bodies — the Hong Kong Chinese General
Chamber of Commerce and the Urban Council — could include a
very small number of rich Chinese who were not British subjects.
There is no convincing reason to believe that these rich Chinese,
many of whom had fled in the face of the advancing Communists
in China in the previous few years, would cast their votes for a
pro-Communist candidate in any election. Nor had the political
situation in Hong Kong in 1952 changed to become more conducive to pro-Communist propaganda.[28] If the experience of the
Urban Council since 1952 is any guide, this so-called 'danger' was
more potential than real. In the final analysis, the Governor
referred to a 'danger' probably because he knew that it would be
the magic word which would induce the Colonial Office to apply
the brake on reform.

In sharp contrast to previous occasions on which a major
change in the direction of the Hong Kong reforms had been
discussed, this occasion did not lead to any controversial discussion in Whitehall. Paskin objected mildly to the volte-face on two
grounds. First, the Secretary of State had already told the people
of Hong Kong during his visit that the reforms were receiving his
sympathetic consideration. Secondly, back-pedalling would require that the Cabinet reverse its approval of the reforms which
the Secretary of State had just secured. Paskin did not, however,
put forward any argument in favour of reform and Grantham's
and the Unofficials' case for abandoning reform was not seriously
questioned. The matter was largely settled when Sir Charles
Jeffries, the Deputy Under-Secretary overseeing Hong Kong
affairs, minuted in support of Grantham. While he conceded that
an abrupt turn-round would be awkward, be preferred to avoid
major constitutional changes during what he called 'the Emergency'. Lyttelton, unlike Creech-Jones (and, to a lesser extent,
Griffiths), was quite prepared to abandon constitutional reform
for Hong Kong, his reason being that the matter did not interest
the British electorate.[29] Although Lyttelton had endorsed Gran-

tham's recommendations, the matter was not referred to the Cabinet for a decision until September because it was thought that the decision should be made public only after Grantham had returned to Hong Kong in October; the delay was not due to any extended discussion or controversy.

In the meantime, the unofficial members of the Urban Council had asked the government to change the Council in the following ways: by increasing the number of directly elected unofficial members from two to four; by extending their term of tenure from one to two years; and by extending the franchise to include British subjects hitherto excluded and all direct taxpayers. With the exception of the extension of tenure, all of these suggestions had already been included in the scheme approved by the Cabinet in May. Instead of seeing this as an indication that the government should proceed with the approved scheme, K. M. A. Barnett, Chairman of the Urban Council (and a senior Cadet), suggested that the government should not proceed to implement all of the approved reform proposals concerning the Urban Council. He requested the government not to reduce the number of ex-officio members in the Urban Council from four to two, his reasons being that the unofficial members of the Urban Council had not made this specific demand and that a large number of ex-officio members were needed to serve on special committees. He also objected to the Urban Council being given greater financial control, on the grounds that it was premature. While the unofficial members of the Urban Council had for some time been considering making proposals for greater financial responsibility, they had not put forward any proposal. In fact, Barnett was opposed to any reform of the Urban Council unless it were absolutely essential to satisfy the expressed views of the unofficial members of the Urban Council. He was also decidedly unhelpful to the newly elected unofficial members of the Urban Council. When Bernacchi tried to move five motions in the Council shortly after he had taken his seat, Barnett promptly suppressed two of them (one dealing with corruption and the other with housing). Barnett pointed out and took advantage of the unpublished standing order that if the Chairman wished to speak, others must be seated and remain silent. His retrogressive recommendations were approved because they were supported by the unofficial members of the Executive Council.[30]

On 18 September 1952, the British Cabinet once again considered the Hong Kong reforms. Lyttelton recommended that the Legislative Council reform be abandoned on the grounds that whilst the situation in Hong Kong had not deteriorated it had not improved, and that it would be politically practicable to abandon reform. He proposed only to increase the number of unofficial members in the existing Urban Council from two to four. Following the wishes of the Executive Council, Lyttelton had left out all of the other suggested changes to the Urban Council which had been agreed earlier in the year (namely, a decrease in the number of ex-officio members from four to two, the widening of the franchise, and expansion of the Council's financial responsibilities). The Cabinet, which was uninterested in the Hong Kong reforms, approved Lyttelton's recommendation without discussion or comment; the British government's pledge of 1946 to give the people of Hong Kong greater local self-government was ignored.[31]

The primary concern of the Executive Council and the Governor at this stage was convincing the people of Hong Kong that they had fought hard against Whitehall to secure the reforms. For this purpose, they requested that the Secretary of State limit his public statement to two points: the 'postponement' of the Legislative Council reform and the increase of elected unofficial members in the Urban Council from two to four. Shortly after Grantham's return to Hong Kong, Lyttelton duly made such a statement in the House of Commons on 20 October, stressing that the time was 'inopportune' for constitutional changes of a major character. To the annoyance of the Colonial Office, and without giving a reason, Grantham disregarded its request that he make a simultaneous statement on the subject in Hong Kong. He made a supplementary statement, apparently sympathetic to reform in the Legislative Council, two days after the Secretary of State had publicly rejected any reform of the Legislative Council. Grantham assured the people of the colony that he was 'at all times ready to consider further proposals for constitutional changes provided they were not of a major character'. He added: 'Indeed, in this connexion a number of recommendations made by the Urban Council itself are at the present receiving the consideration of the Government.'[32] Grantham had succeeded in giving the impression that it was Whitehall, not Government House, that had rejected the notion of giving the people of Hong Kong greater self-government.

Public reaction to this decision was mixed. Sceptics of constitutional reform, such as the *South China Morning Post*, welcomed the announcements as 'wise and appropriate'. Supporters of reform, such as the Reform Club, held that this very limited change was inadequate. While the Club was prepared to take advantage of the announced change to contest all four seats at the Urban Council in 1953, it also declared publicly that there should be at least two further changes: the introduction of at least two elected members to the Legislative Council by 1953; and a further increase in the number of elected members to the Urban Council to at least equal the number of nominated members. The *Sing tao jih pao* acquiesced in the decision, but stressed that a more decisive step should and could have been taken. It emphasized that since the Chinese Communists had taken power on the mainland, the Chinese in Hong Kong had decided to stay; the Chinese residents in the colony were no longer sojourners. In these changed circumstances, the *Sing tao jih pao* argued that the old system of nominated representation of the Chinese community was inadequate and it called for the introduction of elected representatives into both the Legislative and Executive Councils. Although the majority of the other previously articulate public bodies and leading Chinese-language newspapers remained silent, this certainly did not mean that they supported the announcements. In some cases, particularly the right- and left-wing newspapers, their silence probably indicated indifference. They had in any case never really been keen about the reforms, especially since 1949. In other cases, including the majority of the pressure groups, their silence signified acquiescence.[33] Four factors may have contributed to their silence. First, it was not part of the Chinese political tradition to fight for a lost cause. Since the British government had made up its mind, it would have been pointless to reiterate the old demands; they would only have lost 'face' by publicly demanding what they knew would not be granted. Secondly, the depression and the series of minor riots caused by over-zealous Kuomintang agents earlier in the month might have diverted the organizations' attention. Thirdly, the length of time that had passed before the announcement had been made had prepared them to accept something less than the reform proposals discussed in 1949. Finally, the Governor's assurance that the change in the Urban Council was merely the first step leading to further changes had also made Lyttelton's announcement much easier to accept.

The silence of the Chinese Reform Association was presumably due largely to the deportation of its chairman, Mok Ying-kuei, just a month before Lyttelton made his announcement. While Mok was undoubtedly deported for his political activities in connection with the Chinese Communists, it is unclear why he was deported at that time. Although the Executive Council had singled him out as a Communist subversive at the time of the 1 March Riot, it did not order his deportation until September. Whether or not the timing of Mok's deportation and Lyttelton's announcement was purely coincidental, it had the effect of pre-empting the Chinese Reform Association from making any unfavourable public comment. Mok's deportation must have made the other leaders of the Association wary of what they said in public. It had also shattered the Association's political prestige. The Association was also busy at the time selecting a new chairman.[34]

Compared with the increase in the number of unofficial members in the Executive and Legislative Councils in 1948 and 1951 respectively (which had not been given much publicity), this minor change to the Urban Council was publicized out of all proportion to its importance. Described as a 'constitutional development', it was in fact more of an administrative adjustment. After all, despite its name, the Urban Council was not a municipal council but a government agency with a large advisory role, and it could have been expanded without seeking prior permission from the British Cabinet; the government of Hong Kong had only to amend the Urban Council Ordinance.[35] Indeed, the Governor had not sought London's prior approval either for the reintroduction of Urban Council elections in 1952 or for the expansion of the Executive and Legislative Councils in 1948 and 1951 respectively, both of which should be considered to be major reforms under the 1952 definition. However, the wide publicity given to the Urban Council change had the effect of making the abandonment of all major reforms more acceptable to the local people.

Although the Governor and the Executive Council had recommended that the British government restrict the reform to an increase in the number of elected unofficial members in the

Urban Council, they were prepared to accept further changes. They were fully aware of the demands of the unofficial members of the Urban Council, and they also expected some local agitation to follow the October announcement of the abandonment of all major reforms. Consequently, they decided to implement the Urban Council changes in two stages and to present the second stage as a response to public opinion. Shortly after the October announcement, the Hong Kong Reform Club petitioned the government to introduce two elected members into the Legislative Council (this being a very much watered-down version of the Club's demands set out in its petition of 1949). When this demand was turned down on the grounds that any change to the Legislative Council would be a constitutional change of a major character, the Reform Club launched a campaign to gain public support. It collected over 12,000 signatures before submitting its next petition for reform to the Governor in 1953. The Governor responded by dismissing the Reform Club as representing nobody but itself, and by stressing that some of the people who had signed the petition were hawkers and were therefore irresponsible. (This indicates the arrogant way in which Grantham viewed the lower-class Chinese, the silent majority in the colony.) He and the Executive Council also responded by introducing the additional changes to the Urban Council which they had agreed in the winter of 1950 and the summer of 1952. These further changes were primarily related to expansion of the Urban Council franchise to include specific groups of people such as British subjects hitherto excluded because of membership in the Essential Services or ignorance of the English language, and extension of the tenure of unofficial members in the Urban Council from one to two years. Unlike the 1952 change in the Urban Council, these further changes were decided in the Executive Council without seeking prior permission from Whitehall.[36]

While Grantham and the Unofficials had not taken the initiative in asking for the reform proposals to be taken out of cold storage at the beginning of 1952 in order to scrap them, they did not hesitate to do so when they found this to be practicable. Despite Grantham's use of the term 'danger' when he had first asked the Colonial Office to drop the Legislative Council reform, security was not the crucial consideration; no one in the British or Hong Kong governments seriously doubted that the proposals approved by the Cabinet in May 1952 could have worked. The most important factor which caused this turn of events was the attitude of the Governor and of the Unofficials. Two other important contributing factors were the indifference with which the British government had treated the issue, and the lack of widespread public demand for reform in the summer of 1952. When the Unofficials and Grantham saw that they could get away with a mere gesture, they immediately seized the opportunity. The volte-face in the summer of 1952 was quite unplanned; it was an anticlimax.

Economic Depression and Reform

Despite the strong outburst of concern from local businessmen and journalists when the United States imposed strict trade restrictions on Hong Kong in December 1950, the restrictions did not cause a mild recession until the latter part of 1951. The value of trade for the first six months of 1951 in fact accounted for over fifty-eight per cent of the total trade for the year. This was largely responsible for the increase of over nineteen per cent in the value of trade (in current prices) above the 1950 figure, which reached an unprecedented HK$93,000 million. However, this figure is deceptive in that the general price level had risen sharply during the year; the volume of trade had actually fallen by one million tons. Both the monthly value and volume of trade indicate that Hong Kong's trade situation definitely worsened in the latter part of the year. A steady decline began from June onwards. Hong Kong's business community and newspapers responded swiftly and loudly to the American trade restrictions at the beginning of the year for three reasons other than their economic effect. Hong Kong felt insulted, in addition to being injured, by the Americans who had totally disregarded the fact that Hong Kong had already imposed its own control over strategic exports to China since the outbreak of the Korean War. The American decision to off-load goods from ships already on their way to Hong Kong also produced a great deal of confusion and apprehension. Finally, the trading community hoped to reason with the Americans, through the British government, to relax certain restrictions.[37] Hence, despite the fact that Hong Kong was not yet in a depression in early 1951, the news of the American trade restrictions had caught the imagination of the local people, leading to fear of a depression. It was under such circumstances that Grantham decided in February 1951 to put aside the reforms.

If the alarm had been sounded in 1951, then 1952 saw the real break in post-war Hong Kong's continuous trade boom. By 1952, the full impact of the trade restrictions was being felt. In addition to the American restrictions, Hong Kong also suffered from the United Nations' partial embargo on the export of strategic materials to China, imposed in May, and the further tightening of British export control to China in June 1951. For the first time in post-war Hong Kong, the annual value of trade actually declined by over twenty-eight per cent to HK$6,687 million, even lower than the 1950 figure. The volume of trade was reduced by thirteen per cent. The decline was not, however, caused solely by the trade restrictions. Hong Kong's trade with China also suffered seriously from the launching of two political campaigns inside China: the so-called *san-fan* (the three antis) campaign at the end of 1951 against corruption, waste, and bureaucracy; and the *wu-fan* (the five antis) campaign at the beginning of 1952 against tax evasion, bribery, the stealing of state property, scamping on government contracts, and the stealing of state economic information. The legitimate trade between Hong Kong and China almost came to a standstill during these campaigns, and it only began to recover after the campaigns were largely over in May. According to Hong Kong General Chamber of Commerce calculations, the decline in Hong Kong-China trade in 1952 was substantial: in 1950, China accounted for fifty-six per cent of Hong Kong's trade; in 1951, it accounted for forty-nine per cent; but in 1952, the proportion was only thirty-four per cent. These figures do not agree with the statistics prepared by the Hong Kong government, according to which China was only responsible for thirty-one per cent of Hong Kong's trade in 1950, twenty-nine per cent in 1951, and twenty-four per cent in 1952. Nevertheless, the 1952 figure represents a considerable decline in the China trade from the boom of 1950. Hong Kong was also suffering at this time from growing competition from Japan, which had begun to increase significantly its exports to Hong Kong's Asian markets.[38]

Although the economic situation in Hong Kong in 1952 was bad, it was, in Financial Secretary A. G. Clarke's words, 'not by any means desperate'. Two important factors mitigated the effect of the depression. Hong Kong had enjoyed a continuous trade boom for five years since the restoration of civil government in 1946, a remarkable achievement in the post-war Far East. Despite the considerable fall in the value of trade, the total value of trade at current prices in 1952 was still over thirty-one per cent above the 1949 figure, or almost two and half times that of 1947, the first post-war year under civil rule for which a full year's trade return is available. Moreover, the trade figures for 1950 and the first half of 1951 were inflated by speculators' frantic buying following the outbreak of the Korean War. The depression was therefore not as serious as it would appear from a statistical comparison with the preceding two years. Given the speculative nature of the high level of economic activity in 1950 and early 1951, the mild depression was not entirely bad for the economy. The government of Hong Kong itself in some respects welcomed a mild recession, the reasons being that the period of easy and large profits had lasted too long, and that the mild recession had induced business to be developed on a sounder and less speculative basis. By the end of 1952, the great majority of firms surveyed by the Hong Kong General Chamber of Commerce, including practically all of the British firms, had adjusted successfully to the changed circumstances.[39]

During the recession of 1952, small Chinese trading and industrial concerns were harder hit than their British counterparts. In general, British trading establishments — which were on the whole large, less speculative in their investment policy, and in some cases multinational in their operations — were better equipped to endure a recession than were the smaller local Chinese concerns. More specifically, local Chinese firms were hardest hit because they were more involved with the China trade, particularly during the speculative boom of 1950 and early 1951. Since the Chinese Communists' take-over of power in China, Hong Kong's China trade had passed increasingly into the hands of the local Chinese merchants as a result of the Communist government's policy of eliminating the pre-1949 pre-eminent position of foreign (that is, primarily British) traders in China. In contrast to their cautiously optimistic view in 1949 of the China trade, British traders recognized in mid-1951 that their business had reached a crisis point. By 1952, British trading interests had been largely wiped out in China — so much so that even the British Chamber of Commerce in Shanghai and nine of the pre-1949 total of eleven British consulates in China had either already been closed down or were virtually closed.[40]

The fact that the Chinese business community bore the brunt of the depression is particularly important in the context of this study. The original 142 Chinese organizations that had demanded reform in 1949 consisted of, amongst others, chambers of commerce and manufacturers' unions. Members of these chambers and unions were generally the smaller Chinese trading and manufacturing concerns — the group hardest hit by the depression. According to the *Wah kiu yat pao*, in 1952 alone, well over thirty such concerns, including three banks, collapsed, and many hundreds narrowly escaped bankruptcy either by being taken over or by being reorganized.[41] Since their member companies were preoccupied with the struggle to survive the recession, it is hardly surprising that the majority of these organizations remained silent about constitutional changes, even when the decision to abandon reform was announced. The depression provided a favourable environment for the government to abandon any constitutional reform of substance and to push aside its promise of reform.

The Politics of Security

The political situation in Hong Kong in 1951 was, on the whole, calm and stable. However, a series of events at the beginning of the year served to bring public morale to its lowest ebb during the year. In early January, the United States publicly gave the colony a vote of no confidence when its Consul-General advised American citizens to leave the colony, and the American-owned Chase Bank closed its office in Hong Kong. Other American companies also started to reduce the scale of their business and to pay 'danger money' to their employees in the colony. By coincidence, the Hong Kong government later the same month announced its decision to register all British subjects. This announcement proved to be a mistake as far as its effect on public morale was concerned. Although the announcement by the Hong Kong government had nothing to do with the American activities, the local Chinese inevitably connected the two.[42] The local newspapers generally dismissed the American activities as ill-advised and stressed that the colony was safe, while remaining silent about the government announcement. Public morale had suffered.[43] Later in the year, a number of British traders voiced concern over wealthy Chinese increasingly investing their money overseas. However, according to Grantham, this concern was as unnecessary as it was ill-conceived. Following the very large influx of Chinese capital into Hong Kong prior to the Communist take-over in China and the depression in Hong Kong, it was natural that some of the Chinese capital sought investment opportunities abroad. There is no evidence that the people of Hong Kong had become overly concerned with security in 1951.[44]

In general, the situation in the colony in early 1952 was delicate and potentially explosive, but this was not entirely new; it had been the situation since the Communists obtained control in China. This was due basically to the sharp division of political loyalties amongst those people concerned with Chinese politics. The vast majority of the Chinese inhabitants in the colony were, in the words of the Police Commissioner, 'non-political and concerned with obtaining [their] daily round'. However, the small section of the community which was concerned with Chinese politics constituted a potential security problem: the pro-Kuomintang group might clash violently with the pro-Communist section. The colony was also vulnerable to the Chinese Communist government because of its almost complete dependence on the mainland for supplies of food commodities required by the Chinese inhabitants. The beginning of the depression in late 1951 further aggravated the situation. Contrary to the government's expectation, the problem of a larger than usual population, swelled since 1949 by refugees from China, had increased rather than decreased. Up until 1953, the Hong Kong government expected the several hundred thousand refugees to return to the mainland once the Communists had stabilized the situation there. However, unlike Chinese refugees in the past, who had been fleeing from Chinese rebels, warlords, and the Japanese, most of the refugees who fled from the Communists did not return to their home villages or towns once peace had been restored. This contributed to a serious problem of unemployment and underemployment, even during the boom year of 1950. The recession had only made the situation potentially more explosive.[45]

Compared with their behaviour in the past, the Communists were relatively more aggressive in Hong Kong in early 1952.

However, in the latter half of 1952 they resumed their old policy of conciliation towards the colonial government. They had not changed their fundamental attitude and policy towards the colony, and there was no indication that they intended to approach the British with regard to its political status. Their threat to Hong Kong's security remained more potential than immediate. Nevertheless, they constituted the primary security concern of the government of Hong Kong.

Peng Chen, a member of the Communist Party Political Bureau, summed up the Chinese Communist government's attitude towards Hong Kong in the early 1950s when he said:

It is unwise for us to deal with the problem of Hong Kong rashly and without preparation. To take back Hong Kong now would not only bring unnecessary technical difficulty in the enforcement of our international policy but would also increase our burden. It would be of some advantage to our economic reconstruction to let Hong Kong preserve the status quo.[46]

This statement was generally in line with the Chinese policy towards relations with Britain. Sadar K. M. Panikkar, Indian Ambassador to China and the non-Communist foreign diplomat who was closest to the Chinese government, observed that China not only wished its relations with Britain not to deteriorate, but it also wanted to prevent such a thing from happening. At the same time, the Communist leaders were anxious to rid the country of all vestiges of its pre-war 'semi-colonial' status. Consequently, despite its insistence that Britain should prove its sincerity in seeking normal diplomatic relations with China (by severing relations with the Kuomintang government, by voting for representation of the Chinese Communist government at the United Nations, and by protecting Chinese state properties in British territories, particularly Hong Kong), the Communist government had at least up until October 1951 concentrated its propaganda attack mainly on the Americans and had avoided the British as far as possible. Hong Kong was seldom mentioned. Although British treatment of the Chinese in Malaya was occasionally attacked, this was primarily for consumption among overseas Chinese in South-east Asia. The Malayan policy of deporting unwanted Chinese was given as little publicity as possible once the deportees arrived at reception centres inside China.[47]

Shortly after Winston Churchill became Prime Minister in October 1951, the Chinese Communists intensified their propaganda attacks on the British, particularly with reference to Hong Kong. The most important contributing factor was their perception of Churchill's foreign policy. The Chinese Communists had for a short while been mistaken in thinking that, unlike Attlee, Churchill would abandon Britain's independent China policy and would side with the Americans. They had been misled by Churchill's declaration of support for American foreign policy, made at Capitol Hill in January 1952. In such circumstances, they could see no need for restraint in launching propaganda attacks against the British authorities in Hong Kong and elsewhere in order to dissuade the British from teaming up with the Americans. Once they had made up their minds, Hong Kong provided a ready target because its government was at that time pursuing a vigorous policy against local Communist supporters. In addition, the Western embargo against China had greatly reduced the economic value of Hong Kong to China.[48] It is unlikely that the Communists launched a propaganda campaign against Hong Kong in early 1952 primarily because of the actions of the colonial government. Although the Hong Kong government had taken a series of what were considered to be unfriendly actions earlier in the year, the Communists had not at that time launched a propaganda attack against the colony. The most serious of these actions was the requisition of a Chinese tanker, SS *Yung Hao*, in the colony in April 1951. The British Cabinet had acted in this incident under very strong American pressure. The Chinese regarded the incident as a more blatantly unfriendly act than the continuing dispute over the seventy-one aircraft (which lasted from November 1949 to July 1952), because the tanker was not a warship and its ownership was undisputably Chinese. They protested strongly and retaliated by requisitioning all the properties (with the exception of the offices) of the Shell Company inside China.[49]

The Chinese Communist propaganda campaign against the Hong Kong government began in January 1952 with its focus on what they called the persecution of Chinese in Hong Kong (specifically, the deportation of film workers, union leaders, and other pro-Communist activists). Although none of those people were taken to court, the government believed that it had good grounds for proving that they were connected with the Communists or were infiltrating the film industry, or both. The second theme of the initial campaign was that the Hong Kong government was helping the Kuomintang to infiltrate China by turning a blind eye to the guerrilla activities of the Kuomintang near the border. If the specific guerrilla activities referred to by the Communists had taken place, they did so without the knowledge of the Hong Kong government. From March, three further subjects were added to the propaganda attack: the Hong Kong government's handling of a riot which occurred on 1 March and which caused the death of a Chinese; the government's suspension and prosecution of the publishers and editors of the three local Communist newspapers for publishing an editorial by the *People's Daily* (*Jen min jih pao*) about the riot, which the government of Hong Kong had deemed to be seditious; and the government's decision to close the Hsin Hua News Agency unless it became registered.[50]

Despite the fact that this propaganda campaign was unprecedented, what really distinguished it was the relative restraint exercised by the Chinese Communists. First and foremost, with the exception of one editorial by a Canton newspaper, the *Lien ho pao*, the propaganda was restricted to specific acts of the colonial government. The *Lien ho pao* was particularly critical because it thought that both the American and British 'imperialists' were using Hong Kong for aggression against the Chinese people. It warned that Hong Kong would 'one day be liberated from the savage and despotic rule of the imperialists'. This warning was not as threatening as it might have been, however, since it did not commit the Chinese state to liberating the colony, nor did it give a timetable. Whether or not the *Lien ho pao* was aware of the fact, Hong Kong could also be 'liberated' by democratic reform or by the achievement of dominion status.[51] This editorial — intended more for Cantonese than for Hong Kong consumption — also illustrates the fact that propaganda originating in Canton was on the whole much harsher in its tone and content than that originating in Peking or Hong Kong. This was probably due to the central government's concern not to damage Sino-British relations in the long term, while the Canton authority was more interested in blackening Hong Kong's image in order to discourage the Cantonese from trying to escape to the nearby capitalist enclave. For obvious reasons, the Hong Kong Communists had to watch what they said. In general, the Communists had also chosen not to emphasize the importance of the riot, as well as the suppression of their newspapers and of the Hsin Hua News Agency — issues which were of far greater importance to them than the earlier deportations, which had been happening on a lesser scale for some time. Had the Communists chosen to harden the tone of their propaganda in accordance with the importance of the subject matter, they would have been very much more vicious and virulent since March. In any case, the Communists wound up their campaign towards the end of May. The most likely reason for this was that by then they understood that despite Churchill's wartime connection with the Kuomintang government and his Capitol Hill declaration, British Far Eastern policy had remained unchanged. In the final analysis, the propaganda campaign of early 1952 was an aberration; the Chinese Communists had not changed their fundamental attitude and policy towards Hong Kong and Britain.[52]

In the context of this study, the timing of the beginning and end of the propaganda campaign is particularly important. When Grantham wrote in January to ask the Colonial Office to reopen the case for reform, the campaign had just begun. It continued for as long as the reforms were being examined in London. When the British Cabinet approved the reforms on 20 May, it did not know that the campaign was being wound up. When the Unofficials and the Governor decided in June to ask the Cabinet to reverse its approval of the reforms, the campaign had ended. Both the decisions to reform and not to reform were made without regard to the progaganda campaign, even though it could have been seen as a Communist challenge to the government.

The riot of 1 March was described by the Commissioner of Police as a second 'trial of strength between the Government and the Left-Wing trade unions', but this was an overstatement. The origin of the incident was a fire on 21 November 1951 in the squatter area of Tung Tau, near the Walled City of Kowloon, which destroyed 5,000 huts and rendered over 20,000 people homeless. In the aftermath of the fire, a few trade union activists joined the fire victims to organize relief for the victims, but the union activists were deported in January. In the meantime, various organizations in Canton also collected money for the victims. In February, the Canton organizations enquired through Ko Cheuk-hung, Chairman of the Hong Kong Chinese General

Chamber of Commerce, whether the Hong Kong government would permit a Canton delegation to visit the fire victims and distribute relief money. The government refused permission. Towards the end of February, Percy Chen again requested permission for a Canton delegation to visit the fire victims, under a gentleman's agreement that the delegation would not come as a 'Comfort Mission' but as representatives of the donors and would not make political speeches in public. This time, permission was granted. The Canton organizations had avoided relying on the Hong Kong Federation of Trade Unions as its channel of communication with the Hong Kong government, even though the Federation was as involved in the relief work as were the Hong Kong Chinese General Chamber of Commerce and Chen. While the motives of the few unionists who became involved and were deported cannot be ascertained, there is no doubt that the charity of Canton was aimed at publicity and propaganda. It is most unlikely, however, that the sponsors of the delegation in Hong Kong intended to use the visit to create an incident. Ko, Chen, and others undoubtedly knew that they would be deported if anything were to go wrong, and deportation was the last thing they would have wanted. The reactions of the Canton and Peking authorities after the event indicate that they, too, could not have planned an incident. The delegation was scheduled to arrive on the afternoon of 1 March but, for reasons not entirely clear, it did not manage to cross the border that morning. The delegation informed the Secretary-General of the Hong Kong Chinese General Chamber of Commerce of the delay before noon. Although an announcement was made to the press, the news did not reach many of the local inhabitants and a crowd of 10,000 gathered at the Tsimshatsui railway terminal to greet the delegation in the afternoon. When, at three o'clock, the crowd was informed of the postponement of the visit, it left the terminal in an orderly fashion. There were no Communist or Federation of Trade Union agents provoking unrest. When a large group of the departing people reached the Jordan Road junction, half a mile from the terminal, an incident occurred. While the accessible government reports do not explain how the incident began, eyewitness accounts in the *Ta kung pao* indicate that a police vehicle inadvertently ran into the crowd, injuring a girl. The police officers involved exchanged abusive shouts with the crowd and confusion followed. Armed police reinforcements arrived very shortly afterwards, and tear gas was used. The riot had begun. When the dispersing crowd moved a further half-mile north, to a location near the Mongkok Police Station, a police officer found himself surrounded. He opened fire on the crowd, causing one fatal and two non-fatal injuries. Order was restored a little more than two hours after the first clash had occurred. The riot was not planned; it occurred as a result of an unfortunate accident in a tense atmosphere. There is no evidence that the Federation had intended it to be a trial of strength with the government.[53]

The riot, like the propaganda campaign, was irrelevant to the decisions made about reform. The incident occurred more than two months before the Cabinet first decided to approve the reforms. Had the Unofficials or the Governor deemed it a sign of danger in regard to the reforms, they would have prevented them from being submitted to the Cabinet. Indeed, had the Unofficials and the Governor shared the view of the Commissioner of Police that the riot was a challenge to the government, it would have meant that they did not consider a Communist challenge an adequate reason for stopping the reforms. In any case, there is no evidence to suggest that the Governor had linked the riot to the reforms.

The Chinese Communists resumed a policy of moderation towards Hong Kong in the latter half of 1952. The clearest indication of this policy was the self-restraint with which they handled two potentially very explosive events: the final disposal of the seventy-one Chinese aircraft in the colony, and a series of Kuomintang attacks against the Communists in October. In mid-July, the government of Hong Kong had begun making preparations to cope with possible unrest as a result of the expected Privy Council ruling in favour of the Americans on the ownership of the Chinese aircraft grounded in the colony since the end of 1949. The Privy Council announced its decision at the end of the month, but the expected Communist retaliation did not materialize. In the eyes of the Chinese, insult must have been added to injury when the United States Navy sent an aircraft carrier to remove the aircraft. This notwithstanding, the pro-Communist employees of both airlines — the Chinese National Aviation Corporation and the Central Air Transport Corporation — merely sent two letters of protest to the government, and local Communist newspapers reported the event without comment. The precautions of the Hong Kong government proved to be unnecessary because the Communists had decided to meekly acquiesce in the loss of the aircraft.[54]

The attacks in October against the Communists by pro-Kuomintang elements were planned and organized by a group known as the 'Anti-Communist Anti-Russian Youth League'. The League was under the directorship of Pang Chiu-kit and had its headquarters in the Kuomintang stronghold of Rennie's Mill refugee camp. The targets of the series of attacks were the offices and personnel of pro-Communist unions, newspapers, and other organizations (including the Hsin Hua News Agency). The first violent attacks occurred in the early morning of 1 October, the Communist Chinese national day. Further assaults occurred on 3 October and a large-scale attack was planned for 10 October, the Kuomintang Chinese national day. None of the attacks caused much disturbance, partly due to swift and effective police action. The League's plan for 10 October was also disrupted by the police, who arrested thirty-two of its leading members before that day. Nevertheless, the League still managed to launch twenty assaults against various Communist targets. The attacks were also prevented from developing into large scale violent confrontations because of the remarkable restraint shown by the Communists. According to the Political Adviser (G. W. Aldington), there was no doubt at the time that the Communists were under strict orders to avoid violent confrontation with their attackers. The fact that the Political Adviser made this point categorically suggests that the police had been alerted in advance by the Communists, who relied primarily on the police for protection during the attacks. The local Communists had resumed their old policy of abiding by the law of the colony in their activities.[55]

If the October incidents showed the Communists at their best in their behaviour in the colony, they also showed the Kuomintang at its worst. The attacks occurred for two main reasons. The League was probably encouraged by the recent set-backs of the Communists, such as the aircraft incident and the deportation of Communists by the government. Moreover, Peng and his lieutenants may have tried to ingratiate themselves with the Taiwan authorities before their League was amalgamated into General Chiang Ching-kuo's 'Anti-Communist Nationalist Salvation Youth Corps'. It is unlikely that the attackers turned to violence because of their intense hatred of the Communists; most of the attackers were paid to do so. The attacks were also neither ordered nor approved of by the Kuomintang authorities in Taiwan, who were privately apologetic to the government of Hong Kong after the event. Contrary to the fear of the British Consul in Tamsui, Taiwan, the Kuomintang authorities not only did not make a fuss about the large-scale deportation of those League members who had been arrested, but they positively co-operated in order to facilitate their quiet deportation.[56]

Although the attacks represented the climax of the activities of the Kuomintang in Hong Kong during this period, they also illustrated the weakness of the Party. The fact that the attackers had to be paid to do what the Kuomintang would call 'patriotic duty' demonstrates how little influence and support the Party had in the colony. However little influence the Kuomintang might have had in Hong Kong as a whole, it nevertheless had undisputable sway in the Rennie's Mill area. Most of the other Kuomintang agents arrested in the colony during this period, either for the possession of arms or for criminal offences, were either residents of that area or were closely connected with the people there. This heavy concentration of pro-Kuomintang elements in the Rennie's Mill area had made it a potential *imperium in imperio*.[57]

The other preoccupations of the Kuomintang during this period were the support of guerrilla activities in China and the strengthening of its position in the local labour movement. It made little headway in either area. With regard to the labour field, it concentrated on sponsoring what it called the 'free unions' — not to be confused with the independent unions existing in the colony. Whatever success it had in this field had as much to do with the workers' disillusionment with the pro-Communist Hong Kong Federation of Trade Unions as with its own efforts. In any case, it did not constitute a security problem for Hong Kong.[58]

The primary security concern of the Hong Kong government in the early 1950s was undoubtedly the activities of the Chinese Communists. This being so, over-zealous Kuomintang supporters actually caused more violent disturbances for political reasons than did the Communists. (In sharp contrast, despite their rising influence and increasing level of activities, the Communists did not create any violent incident until the Cultural Revolution in the 1960s.) This was not because the Kuomintang was hostile to the Hong Kong government or wanted to destabilize the colony, but because, compared with the Communist Party, it had less strict control over its supporters, some of whom were members of the Triad society. Notwithstanding this, the Commissioner of Police was justified in the 1950s in considering the Communists more dangerous than the Kuomintang.[59] Despite the relatively more volatile political situation in Hong Kong in 1952, it was not directly responsible for the government's decisions first to approve and then to abandon constitutional reform.

7 A Bird's-eye View

A Comparison of the Various Reform Proposals

Between 1945 and 1952, five sets of proposals to reform the Hong Kong constitution were examined, and four of them — those of G. E. J. (subsequently Sir Edward) Gent of the Colonial Office (1945), Sir Mark Young (1947), the Unofficials (1949), and Sir Alexander Grantham (1950) — were rejected by the British government. The fifth and final plan, implemented in 1952, dealt only with the Urban Council. Like the three sets of proposals examined by the Colonial Office after 1949, the Urban Council scheme was sponsored jointly by the Unofficials and Grantham. The only difference between the proposals in this respect was that the Unofficials had publicly identified themselves with the 1949 alternative, whereas Grantham undoubtedly played the leading role in putting forward the 1950 plan. The five proposals are compared, together with the actual constitutional structure when civil government was restored in May 1946, in Table 7.1.

According to the schemes of Gent and Young, the main emphasis of reform before 1949 was the introduction of municipal self-government. Gent and Young also felt, however, that the colonial legislature should be changed to complement their municipal schemes. From 1949 onwards, the main emphasis in the reform proposals shifted to reform of the Legislative Council; Grantham's suggestions about the Urban Council were of only minor importance. Moreover, the notion of introducing a municipal council, or of developing the Urban Council into a full municipal council, was virtually dropped. In late 1952, the emphasis again changed: only the Urban Council would be reformed.

The original proposals of Gent and Young amounted to the setting up of a diarchy in Hong Kong. Young's municipal scheme was essentially a development of Gent's ideas. If it had been successfully implemented, it could have had far-reaching effects. Notwithstanding its name, the municipal council outlined by Young was not simply a municipal council on the English model. His proposed Hong Kong Municipal Council was to be a 'super-municipal council', responsible for virtually all of the built-up

Table 7.1 Comparative Composition of the Various Councils in the Reform Proposals of 1946–52

Council	Actual	Gent	Young	Unofficials	Grantham	Approved
Legislative Council	17	14	15	17	16	No change
Directly elected				6		
Indirectly elected	2	7	4		6	
Appointed	6		4	5	5	
Total Unofficial	8	7	8	11	11	
Official	9	7	7	6	5	
Unofficial majority	−1	0	1	5	6	
Municipal Council		23	30			
Directly elected		9	20			
Indirectly elected		7	10			
Appointed		7				
Total Unofficial		23	30			
Official						
Unofficial majority		23	30			
Urban Council	13			To be	13	15
Directly elected	2*			examined by	4	4
Indirectly elected				new Legis-		
Appointed	6			lative	6	6
Total Unofficial	8			Council	10	10
Official	5+				3+	5+
Unofficial majority	3				7	5

Notes: * In abeyance, 1946–51   + Including chairman

area of the colony and with a degree of administrative and financial autonomy unknown in other British municipal councils. If the council could successfully have discharged all of the duties Young proposed to give it, and if the British government had so decided at a later date, it could gradually and eventually have replaced the existing colonial legislature and also have assumed a considerable measure of executive responsibility, thus completely changing the constitutional structure of the colony.

Gent and Young differed mainly with respect to their secondary proposals for the Legislative Council. Gent preferred to let responsible organizations nominate all of the unofficial members of the Council, while avoiding the creation of an Unofficial majority, whereas Young favoured the retention of appointed unofficial members and the creation of a token Unofficial majority of one. This difference was not as great as it may appear, however, because both Gent and Young were prepared to make adjustments in the light of the municipal experiment.

The Young Plan did not reflect exactly Young's original proposals. Even if the Plan as it appeared in the form of the draft bills laid before the Legislative Council in the summer of 1949 had been successfully implemented, it would not necessarily have had consequences as far-reaching as outlined above. In the first place, the British government had decided not to give the municipal council the extraordinary degree of autonomy that Young proposed. Instead of being a 'super-municipal council', it was to have been a municipal council similar to its English equivalent. Secondly, Young's successor did not share his vision of the proposed 'super-municipal council' being used as the means of educating a new generation of Hong Kong citizens who would strive to turn the colony into a kind of a city-state within the British Empire-Commonwealth. Hence, he would not have helped to develop Young's proposed municipal council into a potential alternative government. In other words, even if the Young Plan as such had been implemented, it would have meant only the introduction of an elected municipal council, considerable expansion of the franchise, and the introduction of an Unofficial majority into the Legislative Council. Whether or not there would have been further constitutional development would have depended mainly on the policy makers involved and the response of the local population.

The Unofficials' alternative and Grantham's 1950 proposals also had potentially very important consequences, since they provided for the introduction of a large Unofficial majority. In principle, if either of these schemes had been successfully implemented, it would have prepared Hong Kong for further constitutional advance following the more usual pattern in British colonies — that is, the gradual expansion of the Unofficial majority, together with the appointment of a Speaker to replace the Governor as Chairman of the Legislative Council. There may even have followed gradual evolution towards ministerial government once individual Unofficials had been appointed to oversee specific government departments. Nevertheless, whether or not there would have been further reform is a different matter.

The Unofficials' alternative also had potential implications even more far-reaching than Grantham's scheme because it envisaged the introduction of direct elections to all British subjects in the colony. Although this provision could not have enfranchised (in Secretary of State Creech-Jones' estimate) more than 17,000 people at that time, the situation would have altered completely within one or two generations. Before 1950, most of the Chinese inhabitants of the colony were not British subjects, whereas less than thirty years later, the vast majority of the local population were British nationals. Therefore, had the Unofficials' alternative been implemented, within about thirty years the franchise would automatically have expanded to include the vast majority of the local inhabitants — a development which most of the Unofficials did not believe would occur — even if no other steps for constitutional advance had subsequently been taken.

Unlike its predecessors, each of which had important implications in some way, the Urban Council scheme was of little real significance; the Council was neither a municipal council, nor a centre of power, nor fertile ground for an experiment in democracy. Despite the Hong Kong government's claim that the changes introduced in 1952 represented constitutional advance, they were in fact little more than administrative changes, since the Council functioned primarily as a government agency, and those changes were introduced without amendment to the constitutional instruments of the colony.

The Attitudes and Approaches of the Governors

The most important factor which altered the direction of the Hong Kong reforms during this period was the difference in attitude and approach of the two Governors involved — Young and Grantham. This difference was largely responsible for the turn of events between July 1947 and June 1949. However different their attitudes and approaches, there is no doubt that both firmly believed that their respective approaches served the best interests of the colony. There is no question of either Young or Grantham having made a decision with regard to the reform proposals in order to protect or enhance his own power or interests; they chose dissimilar approaches to reform of the constitution of Hong Kong simply because they had very different views on the subject.

Young fully supported his plan because he thought that given the Chinese government's determination to recover Hong Kong, and the scale of Kuomintang activities inside the colony, the only way to keep the colony British was to make the local inhabitants want to do so. This could only be achieved by making the local inhabitants citizens of (not merely Chinese sojourners in) British Hong Kong through popular political participation. However, in order to do so, and in order to carry out the British government's promise to promote greater local self-government for the local inhabitants, an appropriate set of reform proposals ought to fulfil two further requirements. First, it must be an acceptable alternative to the introduction of popular elections to the Legislative Council because over ninety-seven per cent of the local population were non-British and it was not the usual practice at that time to allow non-British to participate in colonial legislatures. Secondly, since this reform was an experiment which inevitably carried the risk of failure, it was preferable, if not essential, should it fail, that the government could discontinue or put aside the experiment without causing excessively undesirable effects on the constitutional structure of the existing government. It was for these purposes that Young put forward his 'super-municipal council' scheme, which was modestly styled an attempt to introduce a municipal council with suitable changes in the Legislative Council — one which, if successfully implemented, could later be transformed into a potential alternative government.[1]

Grantham not only did not share Young's view, but felt strongly that it was unrealistic to expect Hong Kong to follow the example of other British dependent territories and achieve 'independence'.[2] He believed that Hong Kong had to be either a British colony or a part of the Chinese province of Kwangtung. He believed that the British would have to return the whole colony to China when the lease for the New Territories expired in 1997, and that the form of government most appropriate in the existing circumstances was a 'benevolent autocracy' — that is, the particular version of the Crown colony system then existing in the territory.[3] Consequently, unlike Young who pushed Whitehall to implement his proposals as soon as possible, Grantham saw no need for speed or change. Grantham at no time advocated to the Colonial Office any reform which might change the nature of the existing government for its own sake. The various proposals he put forward during this period (which, compared with Young's scheme, were relatively more controllable by the colonial government, less complicated, and less expensive) were intended either to supersede the Young Plan or simply to make good the British government's promise of reform made on 1 May 1946.[4]

Young and Grantham took such dissimilar views largely because they had different understandings of the Chinese people. This was in turn affected by their previous experiences with the Chinese. As Alastair Todd, Young's private secretary, recalled, Young did not have much knowledge of the Chinese people. This is hardly surprising since he had had no previous experience of the Chinese before he became Governor of Hong Kong in September 1941, and he became a prisoner-of-war less than four months later.[5] As a result, Young probably thought of the Chinese as similar to some of the culturally more advanced colonial peoples in the British Empire, such as the Sinhalese in Ceylon — hence his proposals to use a 'super-municipal council' to encourage the development of a Hong Kong citizenry. In sharp contrast, Grantham had had extensive experience with the local Chinese when he was a junior Cadet in Hong Kong between 1922 and 1935. Grantham justifiably thought that he understood the local Chinese well; he considered them to be unlike other colonial peoples who could gradually develop identities of their own and eventually evolve into independent countries within the Empire-Commonwealth. In his view, the cultural affinity of the Chinese was too strong and Hong Kong was too close to China proper for the majority of the local Chinese inhabitants to develop 'local loyalty' to the colony, let alone allegiance to the British Empire.[6]

The careers of both Young and Grantham before they became Governors of Hong Kong also contributed to their divergent approaches. Young had served in Ceylon, Sierra Leone, Palestine, Barbados, Trinidad and Tobago, and Tanganyika before taking up his Hong Kong appointment. He had a professional training which included working with elected representatives and guiding the relatively more civilized colonies to constitutional advancement. He had begun his career in Ceylon, 'the pioneer of the non-European dependencies' in constitutional development, just before it became the first Asian British colony to have elected non-European representation in 1910. He subsequently spent his formative years as a colonial civil servant in Ceylon until 1928, interrupted only by war service.[7] Grantham had had a different training. Although he had worked with elected representatives in Bermuda and Jamaica, none of the colonies in which he served before 1947 — Hong Kong, Bermuda, Jamaica, Nigeria, and Fiji — had had their constitutional structure changed while he held office. The fact that Grantham had begun his career and spent his formative years as a colonial civil servant in Hong Kong was particularly important in shaping his views on the local Chinese and on other problems unique to Hong Kong. The only important political event in Hong Kong between 1922 and 1935 which Grantham recorded in his memoir was the general strike of 1925.[8] Occurring as it did at the height of a nationalist movement in China, the strike was instigated mainly by Chinese nationalists (affiliated with both the Kuomintang and the Chinese Communist Party, which were at that time allies in a united front) as retaliation against the British 'imperialists' for having, in their eyes, massacred Chinese students and workers during a street protest in Shanghai. If this incident left any impression on Grantham — and it would be invidious to assume that it had not since it was the first and only occasion in peacetime on which the colony was almost paralysed — it would have been that most of the local Chinese could never relinquish their loyalty to China and become faithful British subjects.

In retrospect, although Grantham has been proved right in his prediction as to the future of Hong Kong — which further justifies his decision to abandon Young's reform programme — he was wrong in his view that the Hong Kong Chinese would never become good citizens of a British Hong Kong. Moreover, it should also not be assumed that Young was unjustified in putting forward his scheme. The failure to introduce constitutional reform in the 1940s and 1950s — whether in the form of the Young Plan, the Unofficials' alternative, or Grantham's 1950 proposals — had helped to encourage the local Chinese to remain largely apathetic politically. This in turn made it possible for the British and Chinese governments to decide to return the colony to China in 1997, without causing a major controversy in Hong Kong itself. Grantham's policy with regard to constitutional reform had therefore contributed to making good his prediction.

Notwithstanding Young's relatively poor knowledge of the Chinese, his assumption that the local Chinese in Hong Kong could be transformed into loyal citizens of a British Hong Kong was proved accurate. Thirty years after Young had made his recommendations, the majority of the locally born Chinese, after due consideration, had claimed British nationality without official encouragement. It is worth noting that although according to British law all locally born people are British nationals and that, whether or not they claim British nationality, they are entitled to be treated as British nationals of Hong Kong, this was and still is widely misunderstood by or unknown to most Chinese born in the colony. There exists a widespread misconception that ethnic Chinese born in Hong Kong must make a conscious choice to become British nationals because when they apply for their identity cards at the age of eighteen they are asked whether they would like to claim British or Chinese nationality. Given this misconception, a clear majority still 'choose' to claim British nationality, indicating that they do regard themselves as citizens of British Hong Kong.[9] Had Young's scheme been successfully implemented, it would undoubtedly have accelerated the process of creating a British Hong Kong citizenry. Whilst this would certainly have encouraged (as Young would liked to have seen) the Hong Kong citizens of the 1980s to demand retention of the British connection, it remains uncertain as to whether it could have changed the fate of the colony.

The difference in outlook of the Hong Kong- and Ceylon-trained colonial officials was not limited to Young and Grantham. The information available suggests that Hong Kong- and Ceylon-trained officials responsible for Hong Kong affairs had considerable differences of opinion on the question of constitutional reform between 1941 and 1945 when Hong Kong was under Japanese occupation and British officials in London were planning to recover the colony. Within the Stanley Internment Camp itself, senior officials of the Hong Kong government were divided on the subject. F. C. Gimson, who had no previous experience in Hong Kong, had arrived as Colonial Secretary only one day prior to the outbreak of the Pacific War. He had extensive experience in Ceylon, where universal suffrage and (in Martin Wight's terms) a 'semi-responsible unicameral elected legislature' had already been introduced in 1931. Gimson was at first shocked by senior Hong Kong officials' 'professed ignorance of the policy of the imperial government that the local population of the colonies should be trained to assume authority later to be granted self-government'.[10] R. A. C. North, the second senior official at Stanley (who was a Hong Kong Cadet and at that time Secretary for Chinese Affairs), did not hesitate to contradict Gimson publicly and openly on this issue. The most important underlying cause of this difference, as Gimson rightly saw, was the completely different experiences of officials trained in Ceylon and in Hong Kong. In Ceylon, officials were accustomed to working with the elected representatives and they regarded the welfare of both the Europeans and the Ceylonese equally as being the concern of the government. In Hong Kong, however, officials were accustomed to dealing with the European and Chinese residents as two completely separate communities and overlooking their obligations to the latter.[11]

The views of N. L. Smith (Gimson's immediate predecessor and a Hong Kong Cadet) on the need for divergent approaches to constitutional development in Hong Kong and Ceylon were probably also shared by many of his Cadet colleagues.[12] In London, Smith explained the matter to the Colonial Office when, in the summer of 1942, *The Times* criticized the pre-war Hong Kong government. Smith defended the lack of any constitutional advance in Hong Kong since the turn of the century on two grounds. In the first place, he wrote, 'once the initial seizure by force of arms of this barren rock [was] justified, it was entirely logical to offer immigrants, and even the descendants of such immigrants, the option of coming under the existing constitution or of staying in (or returning to) their own country. In this respect we differed fundamentally from colonies such as Ceylon where forcible imposition of an alien rule on an autochthonous population makes the gradual fostering of self-government almost as necessary as in mandated territories.' Secondly, there was no demand for constitutional advance.[13] Of importance in this context is that Smith (and possibly most of his Cadet colleagues) formed their attitudes not only because they did not have experience similar to that of their colleagues in Ceylon, but also because they considered it right and proper that Hong Kong should not have constitutional reform. In their view, Hong Kong was a very special case where British colonial policy designed for such colonies as Ceylon should not be applied.

Although Gimson was probably correct on the whole in his assessment of the Hong Kong officials, not all of the Hong Kong-trained officials rejected reform during this period. David MacDougall, who headed both the Hong Kong Planning Unit and the Civil Affairs Unit before becoming the first post-war Colonial Secretary, was also a Hong Kong Cadet but he did not share the views of North; he favoured reform. However, MacDougall, in more ways than one, was an exception rather than the rule. He had always admired the Chinese (or, to be precise, the Cantonese, who make up the overwhelming majority of the

Chinese in the colony). In addition, he had had experiences that most of his Cadet colleagues had never had. After escaping from Hong Kong on the day the colony surrendered to the Japanese, he returned to Britain through China to assist in Gent's defence against the Foreign Office's demand in June 1942 that Britain should give up Hong Kong in order to protect other more important British interests. Subsequently, at the end of 1942 and in 1943, MacDougall served in various capacities in New York, Washington, and San Francisco, where he again defended British colonial policies in general and Hong Kong in particular. He further defended Britain's position privately during the Pacific Relations Conference in Mount Tremblant near Montreal, where the Chinese participants said that they wanted the return of Hong Kong.[14] It is hardly surprising, therefore, that MacDougall shared Gent's concern that (both for the benefit of the local people and for the sake of silencing Britain's American critics) Hong Kong's post-war constitution be revised on a liberal basis.

Whitehall and Reform

Between 1945 and 1952, the Colonial Office changed from actively advocating a policy of reform of the constitution of Hong Kong to passively endorsing a policy of avoidance of any major change. Initially, under the direction of Gent and with the assistance of MacDougall's Hong Kong Planning Unit, and in the absence of the Hong Kong government, the Colonial Office played the leading role in initiating and promoting a liberal reform for post-war Hong Kong. The British government's pledge to grant the residents of Hong Kong greater local self-government, made by Young and Secretary of State Hall simultaneously on 1 May 1946, represented the view of the Colonial Office on the subject at that time.

Following the restoration of civil government in Hong Kong, the Governor took over (as a matter of course) from the Colonial Office the task of drafting a plan to implement the British government's policy of reform. The Colonial Office's contribution consisted mainly of 'satisfying itself' that the Governor's recommendations were appropriate for the colony and generally in line with British policy. Although the Secretary of State, with the help of the Colonial Office (his secretariat), was officially responsible for charting the course of development, in practice the Governor in Hong Kong decided the direction of the proposed reforms.[15] On the whole, the Colonial Office continued to support reform in Hong Kong until at least the end of 1948, notwithstanding the fact that for a very brief period during the winter of 1946–7 it was a little hesitant about introducing Young's proposed reforms because only then had it fully recognized the risk inherent in the post-war situation — the possibility of infiltration of an elected council by the Kuomintang.

Although the role of the Colonial Office had not changed, from 1949 it altered its views on reform. It began to be sceptical about the policy of reform of the Hong Kong constitution, the crucial factors being the success of the Communists in the Chinese civil war and a change in the reform proposals. As far as the Colonial Office was concerned, the emergence of a Communist government in China had brought about what it called an 'emergency' in Hong Kong, which made it prudent to reconsider whether or not the Young Plan should be introduced. It is likely, but by no means certain, that the Colonial Office would have acquiesced had Grantham insisted on the implementation of the Young Plan in 1949. However, Grantham's recommendation that the Young Plan be replaced by the Unofficials' alternative, which asked for the introduction of both direct elections and a large Unofficial majority at the colonial government level during the 'emergency', certainly made the Colonial Office turn against reform. The situation was exacerbated by the change of personnel in charge of Hong Kong affairs in the Office. By then, with the exception of Paskin, all of the officials who had previously favoured a policy of reform — Gent, Lloyd, Caine, and A. M. Ruston (now Lady Paskin) — had been replaced by officials who had never backed the policy — Jeffries, Sidebotham, W. I. J. Wallace, and H. P. Hall. From 1950, the Colonial Office continued to examine the various reform proposals primarily because the British government had pledged to do so.

By 1952, the Colonial Office had not only completely changed its attitude on reform of the Hong Kong constitution, but it had also decided to play a largely passive role. The underlying reason for this transformation was that responsible officials in the Colonial Office had by that time become either indifferent, or in a few cases sceptical, about proceeding with the reforms. Hence, the

Colonial Office was inclined to follow the advice of the man on the spot.[16] In any case, the responsible officials were also preoccupied with other more pressing issues, particularly the emergency in Malaya.[17] Consequently, when Grantham asked in the summer of 1952 that the Cabinet's earlier decision be reversed and that changes be limited to the Urban Council, the Colonial Office chose to allow Grantham himself to persuade the Secretary of State. In the light of what occurred in the preceding few years, it is reasonable to conclude that had all the responsible officials in the Colonial Office either strongly supported or opposed reform in 1952, they could have delayed (or even prevented) either Grantham's 1950 proposals from obtaining Cabinet approval in May or the subsequent decision in September to back-pedal.

In total, four Secretaries of State (three Labour and one Conservative) were involved between 1945 and 1952. Of the four, only Creech-Jones was committed to the introduction of a liberal constitution in the colony. Unlike his predecessor and his successors, Creech-Jones was anxious that Hong Kong should also follow the general colonial policy which he himself had had a leading role in formulating: 'dependent territories shall be guided to responsible self-government within the Commonwealth in conditions that ensure to the people concerned both a fair standard of living and freedom from oppression from any quarter.'[18] He also shared Young's view and firmly believed that given the Chinese government's determination to recover the territory, democratic constitutional reform was the only way to ensure that Hong Kong would remain within the Empire-Commonwealth. As a result, he not only approved Young's proposals and attempted to persuade the Colonial Office to implement the Unofficials' alternative at the beginning of 1951 when the Office was dragging its feet, but he also suggested in 1947 that, if possible, ministerial government should gradually be introduced into the colony. Nevertheless, despite Creech-Jones' commitment, he was unable to ensure the speedy implementation of either the Young Plan or the Unofficials' alternative because he was preoccupied with other important and more pressing issues such as Palestine and the general election.

With the exception of Griffiths (Creech-Jones' immediate successor, who had 'very little practical knowledge of colonies' and who supported Creech-Jones' policies), no Secretary of State paid serious attention to the Hong Kong reforms.[19] Griffiths himself was not personally interested. Unlike Creech-Jones (who studied the Hong Kong problem with care, wrote long minutes defining what he wished to do in the colony, and repeatedly pushed the Office to expedite matters), Griffiths rejected Grantham's 1950 proposals purely because he found then to be 'retrogressive' and out of line with the general colonial policy. The other Labour Secretary, Hall, was only nominally involved because he was preoccupied with Palestine during his eleven months in office and the proposed Hong Kong reforms were at that stage non-controversial.[20] Lyttelton, the only Conservative Secretary concerned, also had no special interest in the subject. He left the matter to his official advisers, being relatively less bound than his Labour predecessors by the Labour government's pledge to reform the Hong Kong constitution on a liberal basis. He was also probably unaware of the content of this pledge, since his official advisers apparently had not drawn his attention to it.

Although other departments in Whitehall — notably, the Foreign Office, the Treasury, the Commonwealth Relations Office, the Cabinet Office, and the Ministry of Defence — were consulted at various stages about the Hong Kong reforms, only the Foreign Office offered comments which affected the course of events. As explained in Chapter 5, at the end of 1950 the Foreign Office rejected the Colonial Office's request for its concurrence in the introduction of Grantham's 1950 proposals. On no other occasion did the views of the Foreign Office or of the other relevant Whitehall departments (including the Treasury, but excluding the Colonial Office) cause any change in the direction of reform. As explained in Chapter 4, even the five months' delay in the Treasury's reply to the Colonial Office on the question of financial control with regard to the Young Plan, and its insistence that there be strict financial control, had no important practical effect.

The role which the Cabinet played in this matter was purely formal. The Cabinet was not interested in Hong Kong, except during the period of potential trouble in 1949 and 1950. Once events in Hong Kong had ceased to be potentially damaging to Anglo-American and, to a much lesser extent, Anglo-Chinese relations, the Cabinet could not spare time for Hong Kong. By and large, it merely endorsed whatever recommendations the Secretary of State for the Colonies made with regard to the Hong Kong reforms without displaying any real interest.

The Unofficials and Reform

The unofficial members of both the Executive and Legislative Councils have appeared as shadowy figures in this study, even though they were some of the most influential and distinguished citizens of the colony. For the most part, they chose to avoid the limelight and, except during the Legislative Council debate of 1949, they seldom spoke on the subject in public. They took no initiative in asking questions in the Legislative Council about the progress of reform or in airing their opinions to the press during the six years after the Councils were re-established. On the two occasions on which they asked questions in the Legislative Council — in 1949 and 1952 — the questions were drafted under the aegis of the government.

Before July 1947, the Unofficials failed to influence either the direction or the content of the proposed Hong Kong reforms in any important way. So far as can be ascertained, whilst the Unofficials were consulted by Young, there is no evidence that they expressed any strong objection to his draft proposals. Nevertheless, all of the European members of the two Councils — with the possible exception of M. M. Watson, who was the nominee of the Unofficial Justices of the Peace to the Legislative Council — supported the China Association's unsuccessful attempt to lobby against the Young Plan in Whitehall. In view of this support, the Unofficials' unanimous public rejection of the Young Plan in 1949, and the inadequate records of the views of some of the Unofficials, most if not all of them must have opposed the Young Plan privately even during Young's governorship. However, they made no attempt to lobby Young or to make public their views on the subject. They preferred to swim with the tide for two reasons. First, Young was a 'very authoritarian' and 'somewhat intimidating' person.[21] Secondly, some of the Unofficials were 'a little diffident about opposing strongly the views [Young] might express',[22] and Young was deeply committed to his own proposals. This attitude of the Unofficials is understandable since they depended for the most part on the Governor's goodwill for their positions and they did not have as close a relationship with

Young as they later had with Grantham. According to Grantham, even in his time, the Unofficials as a whole were very unwilling to confront the Governor or the Colonial Office.[23]

The Unofficials began to exert influence on the reforms only after Grantham became Governor. The first occasion occurred during the winter of 1947 when the Executive Council (or, more precisely, Grantham, after consulting the Unofficials privately) decided to put aside the draft bills which were required to put the Young Plan into effect. Since they had found a like-minded friend in Grantham, it had become possible, perhaps even preferable, for the Unofficials to deal with the new Governor directly rather than to lobby Whitehall through the China Association in London. From around late 1948, the Unofficials worked hand in glove with Grantham in proposing all of the changes in both the direction and content of the pending Hong Kong reforms.

Almost all of the Unofficials who served during this period were closely connected with the business community. Leo D'Almada E Castro, a barrister-at-law (subsequently a Queen's Counsel), was the only one not to hold a chairmanship of a large firm and to hold only one directorship (that of China Underwriters Limited). The other Unofficial who did not hold a number of directorships was Arthur Morse, but he was Chairman and Chief Manager of the Hongkong and Shanghai Banking Corporation — 'the bank' in the colony. The rest of the Unofficials each held at least several directorships and/or chairmanships of some of Hong Kong's largest business concerns (including public utility, insurance, manufacturing and shipping companies), banks, and *hongs*. For example, amongst other important positions, R. D. Gillespie was Chairman of Imperial Chemical Industries in Hong Kong; D. F. Landale was a director of the princely *hong* (Jardine, Matheson and Company); P. S. Cassidy was a director of John D. Hutchison and Company; C. C. Roberts was a director of Butterfield and Swire (now Swire Pacific) Limited in Hong Kong; M. C. Blaker was *taipan* of Gilman & Company Limited, Hong Kong; M. K. Lo was a director of China Light and Power Company Limited; T. N. Chau was a director of the Hong Kong Electric Company Limited; S. N. Chau was chairman and managing director of the Hong Kong Chinese Bank Limited; and M. M. Watson, the representative of the Unofficial Justices of the Peace, was a director of the Broadcast Relay Service (Hong Kong) Limited.[24] As a result, even though some of the Unofficials (notably Arthur Morse, who served as one of the China Association's confidential advisers to the Colonial Office in 1945) did favour some kind of reform in post-war Hong Kong, it is doubtful whether they were wedded in any personal way to the idea of reform. First and foremost, they represented the interests of the British and local commercial, industrial, and financial communities. On the whole, they could not possibly have been enthusiastic about any of the proposals for constitutional reform since these would inevitably have involved political risks — and risks were anathema to these communities which were busy re-couping their wartime losses.[25]

In addition, although some of the Unofficials must also have shared Grantham's view that major reform was inappropriate for Hong Kong (particularly after the Communist take-over in Chi-na), unlike Grantham they also rejected all of the major reforms in order to protect or enhance their personal interests. Their rationale is explained most succinctly by Claude Burgess, who served as Defence Secretary and Deputy Colonial Secretary dur-ing this period:

For the most part they were the inheritors of an outdated system of patronage and privilege — a system which would certainly be eroded with every step taken towards constitutional change. They therefore had much to lose. On the other hand it was clear that reform was the Secretary of State's wish — and they knew well enough that privilege was based on obedience. But there was another consideration altogether. They were the ordained (and only) representatives of the people of Hong Kong. If, and when, the popular voice took up the cry of reform, they were bound to heed it and to make some representations to the Government — of which they were, indeed, an extra-mural part. But equally, when the public seemed to lose interest in reform, if only for the time being, they were only too happy to relax their pressure on the Government, and to sit back and hope for better times. It was left to the Government officials to take account of, and to adapt their policy to, these shifting currents.[26]

In other words, the Unofficials put forward their 1949 alternative and agreed to the other schemes of reform merely as the price to pay for superseding the Young Plan and for satisfying Whitehall as well as the articulate public. Hence, when they realized that both the situation in Hong Kong and the personalities in Whitehall made it possible for the price to be lowered, as they did discover during the summer of 1952, they — in the spirit of

Hong Kong businessmen, for that is what they were — asked for
and obtained a handsome price reduction.

The Hong Kong People and Reform

The vast majority of the Hong Kong people expressed no opinion
on the various reform proposals put forward during this period.
Most of them were probably too busy earning a living to bother
about the proposals. They also had no reason to feel excited.
With the exception of the Young Plan, none of the proposals
examined by the British government during this period provided
for the enfranchising of the silent majority, who were mostly
non-British Chinese residents. The bulk of this inarticulate mass,
the illiterate or semi-illiterate coolie class, would not have been
affected even had the Young Plan been introduced. The Plan
provided only for the expansion of the franchise to include all
educated and taxpaying long-term residents. Although this would
have included many people who belonged at that time to the
silent majority, they constituted only a relatively small section of
that group. In view of this fact and the attitude of the local
Chinese at that time, the general support which the Young Plan
received from the local press in 1947 should not be dismissed
out-of-hand. As time went on it became clear that, despite the
indifference of the silent majority, there was a not inconsiderable
number of local organizations and people who were politically
conscious and who had actively supported the introduction of
democratic reforms, particularly since 1949. According to M. K.
Lo's estimate, the outburst of articulate opinion during the summer of 1949 represented roughly 10,000 people.[27] By 1953, despite Lyttelton's October 1952 announcement that there would be
no major reform, the number of local people who supported
reform had increased rather than decreased. Over 12,000 people
signed a petition organized by the Reform Club in 1953, asking
for the introduction of two directly elected members to the Legislative Council. This is not a negligible group, considering that
the Unofficials' alternative proposed only to enfranchise less than
17,000 people.[28]

The opinions aired by different organizations in the colony
were given varying weight by the British officials in London and
Hong Kong, with relatively greater importance being attached to
the views of those organizations which were considered to be

more responsible. In the main, opinions expressed by European-dominated organizations, such as the Hong Kong General Chamber of Commerce and the Kowloon Residents' Association, were deemed to be responsible; those voiced by the Reform Club and the Hong Kong Chinese General Chamber of Commerce were 'not irresponsible'; and those articulated by the Chinese Reform Association were not responsible. British officials also paid little attention to the opinions of the smaller local Chinese chambers of commerce and various kinds of predominantly Chinese local associations.[29] With respect to the local newspapers, although the Hong Kong government generally paid a great deal of attention to editorial opinions, it remains doubtful as to whether editorial comments on the various reform proposals were taken very seriously after the autumn of 1947. Grantham (and the Colonial Office) seldom referred to local editorial opinions on the subject, with three notable exceptions: when Grantham recommended the Unofficials' alternative in August 1949; when he proposed his indirect election plan in 1950; and immediately after Lyttelton rejected all of the major reforms in 1952.

Although the Colonial Office had apparently paid greater attention to public opinion before Grantham became Governor, this fact is misleading. Both the Colonial Office and articulate opinion in 1947 happened to support the Young Plan, and Young was able to assure the Office that his proposals had received overwhelming support from the articulate public. On the other hand, articulate opinion after 1949 was in favour of proposals which both the Colonial Office and Grantham did not accept, and it was the usual practice for the Colonial Office to be guided by the views and recommendations of its man on the spot rather than by local opinion, if they were dissimilar. On the whole, while the Colonial Office did take articulate opinion into consideration, it did not regard it as being of crucial importance. Even the decision in early 1952 to take Grantham's 1950 proposals out of cold storage was based primarily on Grantham's recommendations at that time rather than on opinions expressed during Lyttelton's visit to Hong Kong in December 1951.

In Hong Kong itself, the periodic airing of opinions in the local newspapers served to remind Grantham and the Unofficials that there was a vocal minority which strongly supported reform. This fact had made it politically undesirable for them to reject the proposals for major reform between 1949 and 1952 even if they had ever entertained such an idea before June 1952. In general, Grantham did not pay much attention to these expressions of opinion because he thought he knew what the people, the silent majority, wanted and he considered some of the pressure groups — such as the Reform Club — to be misguided, and others — such as the Chinese Reform Association — to be irresponsible, even potentially subversive.[30] Consequently, he did not even bother to remind the Secretary of State to reply to, not just acknowledge, the petitions of the two reform clubs submitted in the summer of 1949. So far as the Unofficials were concerned, since they were the ordained representatives of the colony and since their views differed from most of the extra-legislative opinion, they dismissed such opinion as representing no one but those organizations and individuals who expressed them.

On the whole, articulate opinion did not cause any change to be made in either the direction or the content of the various reform proposals, at least until the summer of 1952. Indeed, even on this occasion, it was the relative absence rather than the presence of such opinions that was important, providing as it did a conducive environment in which the Unofficials and Grantham could attempt a volte-face.

Economic Changes and Reform

By and large, both the Hong Kong and British governments treated economic and political developments in Hong Kong as two separate issues. The only exception was the Treasury's insistence that all constitutional development be consistent with the principle of Treasury control over the colony; however, this ceased in October 1948 when that control ended. The greatest impact of the economic changes on the policy makers' decision with regard to the proposed reforms was during the depression of 1951 and 1952. In early 1951, Grantham found it politically possible to put into cold storage all of the reform proposals because he felt that the expected depression would divert the attention of the local people, including opinions expressed publicly, from the reform proposals. In 1952, the continuation of the depression made the environment more favourable to the abandonment of all major reforms. In contrast, before the depression caught the imagination of the local population, neither economic rehabilitation in the immediate post-war years, nor economic uncertainty caused by the Communist success in the Chinese civil war, had significantly diverted the attention of the articulate public from constitutional reform. Indeed, the strongest outbursts of support for reform occurred in 1947, when economic rehabilitation was still underway, and in 1949, when the future of Hong Kong and its trade with China were most uncertain. In summary, therefore, economic changes in Hong Kong during this period had only a limited effect on the ways in which the reform proposals were handled, particularly before 1951.

Excluding the attitudes of the officials concerned, the important contributing factors to this limited effect were the fact that, except during 1947, the business community was not particularly keen to influence the direction of the proposed reform and it was in any case unsuccessful in all its attempts to do so. Throughout this seven-year period, the business community had made only four attempts to influence political developments. The first two attempts were the bids of the China Association and the Hong Kong Chinese General Chamber of Commerce, in early and late 1947 respectively, to replace the Young Plan with a municipal scheme on the pre-war Shanghai Municipal Council model. The third attempt was the effort of the various local Chinese chambers of commerce to support the Chinese Reform Association's reform proposals and the support of the Hong Kong Chinese General Chamber of Commerce for the Unofficials' alternative during the summer of 1949. The final attempt was the Hong Kong Chinese Chamber's call for reform when its deputation met Lyttelton in December 1951.

'Danger' and Reform

Between 1945 and 1952, policy makers had referred at various stages to several different kinds of danger. These included the dangers of corruption, political infiltration by Chinese political parties, the introduction of Chinese politics into elected councils in the colony, and the so-called 'danger' of 1952 — that of potentially causing embarrassment to the appointed unofficial Legislative Councillors of Chinese origin and, presumably, also to the government. In one way or another, with the exception of the so-called 'danger' of 1952, these risks made various British officials apprehensive, which, in turn, encouraged some of them to drag their feet with regard to the proposed reforms, particularly after late 1948.

The danger that there would be corruption and wholesale jobbery in an elected municipal council was referred to mostly by Young and by Mayle, Head of the Hong Kong Department in the Colonial Office in 1947. Although Grantham did not mention this danger at all, this does not mean that it had lessened after Young retired. Corruption and jobbery on a large scale had always been a serious problem in post-war Hong Kong until the Independent Commission Against Corruption made a determined effort in the 1970s to stamp out organized corruption.[31] There is little doubt, therefore, that there would have been corruption had Young's proposed municipal council been introduced. However, this should not be seen as a problem peculiar to Young's proposals. Given the situation which existed in Hong Kong in the 1940s and 1950s, any council — elected or not — would have had to face a serious problem of corruption if it were to be given real power in day-to-day administration. The crux of the matter was that government functionaries were able to build deep-rooted corruption 'rackets'. Both the government practice of keeping its functionaries in the same positions for long periods of time, and the willingness of the general public to pay bribes, were important contributing factors. It is doubtful whether the introduction of Young's municipal scheme, or for that matter, any of the other reform proposals discussed, would have caused more (or less) corruption.[32] Although there were direct elections to the Urban Council both before and after the war, the Urban Councillors — whether elected or appointed — were generally not corrupt, whereas the administrative arm of the Council, the Urban Services Department, was notoriously corrupt. The point here is that the Urban Councillors had no administrative power and were consequently not part of the corruption 'racket'. In the context of this study, the important point is that the possibility of corruption did not result in any amendment being made to any of the reform proposals.

On the whole, the threat of infiltration into an elected municipal council in Hong Kong by the Kuomintang was over-stated by Young who had relied on an incorrect appreciation of the situation by Megarry, Acting Secretary for Chinese Affairs. There is virtually no evidence that the Kuomintang had planned to do so.

As far as can be determined, the only reference to a Kuomintang attempt to infiltrate the proposed municipal council was made by Hazlerigg to K. O. Roberts-Wray (Legal Adviser to the Secretary of State) in November 1947 after Hazlerigg had retired. Hazlerigg said that he had indirectly learned that 'some men' had introduced 2,000 new members into the Hong Kong Chinese General Chamber of Commerce in order to secure the nomination of a certain person to the chairmanship of the Chamber and as the Chamber's representative to the proposed municipal council — which was provided for in the Young Plan.[33] Hazlerigg did not refer to the Kuomintang or give any name in his letter, but if he had in mind any political organization, it must have been the Kuomintang since there was no other organization (including the Chinese Communist Party) which might have made such an attempt in 1947. If Hazlerigg did have the Kuomintang in mind, he was over-reacting. The incident he referred to was primarily a manifestation of a power struggle within the Hong Kong Chinese General Chamber of Commerce, and the Chamber was not subject to Kuomintang control, despite the fact that it did try to appease the Kuomintang before that party ceased to be the government of China in 1949. It was generally politically independent of both the Kuomintang and the Chinese Communist Party. Since prominent pro-Communist merchants such as Mok Ying-kuei were among the leading members of the Chamber during this period, and Ko Cheuk-hung (who was successfully elected Chairman) was also well-disposed to the Communist Chinese government,[34] it is difficult to see how the Chamber could have been an instrument for Kuomintang infiltration of the proposed municipal council. In any case, Grantham's official reports on Kuomintang activities during this period made no reference to the incident Hazlerigg mentioned; an indication that Grantham did not think it important.[35]

The threat which the Kuomintang posed to Hong Kong became less and less serious as it suffered successive defeats in the Chinese civil war. By early 1950, the Kuomintang had ceased, even in the eyes of the colonial government, to be a major political force in the colony.[36] Indeed, after the Unofficials' alternative was put forward, there was no further reference to the danger of Kuomintang infiltration in local elections. In spite of the fact that one pro-Communist candidate (Percy Chen) ran for the 1952 Urban Council elections, the Kuomintang made no effort to put up a candidate of its own. This notwithstanding, the Kuomintang did have the resources to win a few seats if direct elections with universal suffrage had been introduced. The Kuomintang also remained a potential threat to law and order in the colony since it was organized (albeit not as tightly as the Communists) and was capable of instigating riots and disturbances — as it did in October 1952.

Before the last quarter of 1948, no official in Hong Kong or London mentioned any danger of Communist infiltration in direct elections. They realized that the Chinese Communists had no intention of becoming involved in local elections. Although the actual situation had not changed, the Colonial Office altered its perception of this matter in a policy review held in November 1948. The responsible Colonial Office officials thought that, since the Chinese Communists would presently replace the Kuomintang as the government of China and would probably also take the place of the Kuomintang in the colony, they could also be expected to infiltrate direct elections in Hong Kong.[37] This proved to be the turning-point in the attitude of the Colonial Office. In its view, infiltration by the inefficient, corrupt, and often inept Kuomintang was one thing — one which could be coped with; infiltration by the efficient, well-organized, and generally (in those days) incorruptible Chinese Communists was an entirely different matter — one which ought to be prevented at all costs. The Malayan experience had undoubtedly influenced the thinking of the Colonial Office.

Notwithstanding the apprehension of the Colonial Office, there is no indication that the Chinese Communists contemplated infiltrating elections in Hong Kong, even after 1949. When Percy Chen (who was then Vice-Chairman of the Chinese Reform Association) participated in the 1952 Urban Council elections, he did so as an individual and he emphasized that he was not running on any organization's ticket.[38] Despite Chen's strong connection with the Chinese Communists, it is very likely that he entered the election on his own initiative. Since the Chinese Communists were at that time waging their only propaganda war against the Hong Kong government during the period under review, when Chen registered as an Urban Council candidate his electioneering should have been linked to those propaganda attacks had it been co-ordinated with the Communist Party. However, it was not. Furthermore, according to Eric Chou (a defector from the Communist *Ta kung pao*), senior Communists in Hong Kong after 1949 were 'very nervous' and secretive about their activities; they worked like 'underground workers', 'magnanimous' though the Hong Kong government was.[39] From late 1948, senior Communist cadres also took pains to reassure the Hong Kong government through third parties that their Party did not plan to destabilize the colony in any way. It is difficult to believe, therefore, that the Chinese Communists had any intention of infiltrating elections in Hong Kong.

A related but altogether different danger was the possibility that the Kuomintang and the Chinese Communist Party might turn Hong Kong into a political cockpit. There were two dimensions to this problem. First, there was the general problem that these two parties might engage in political struggles in the colony. In this respect, labour posed a special problem since it was in this field that the division of political loyalty was relatively most pronounced. From 1946, the gap between the right and left wings of the labour movement widened, culminating in the formation of two different, and politically divided, labour federations in 1948. Whilst it is uncertain whether the government could have prevented such a development, Hong Kong's labour policy under Hawkins certainly contributed to this phenomenon. Notwithstanding the efforts of Ken Baker (the Labour Department's Trade Union Adviser), the Labour Office as a whole had taken no initiative before the Young Plan was superseded by the Unofficials' alternative in 1949 in helping to develop healthy trade unionism in the colony, or in helping the local unions to become better prepared to take part in the proceedings of the proposed municipal council. Given this division of the labour movement, both left-wing and right-wing unions had behaved reasonably responsibly. Despite their almost continuous competition to capture the leadership of the labour movement, both the Hong Kong Federation of Trade Unions and the Hong Kong Trade Union Council avoided direct clashes. Their competition was a peaceful one, and they accepted the existence of a parallel independent (though weak) labour movement. The only occasion on which local unions (in this case, the Hong Kong Federation of Trade Unions and the Tramways Union) acted irresponsibly was during the Tramways strike of 1950; however, the government and the management of the Tramways Company were equally responsible for the excesses of this incident. In any event, whether or not the

Hong Kong Federation of Trade Unions planned to use this strike to discredit the Hong Kong Trade Union Council, the incident did not turn out to be a contest of power between these two organizations.

Secondly, there was the specific problem that the Kuomintang and the Chinese Communist Party might introduce Chinese politics into Hong Kong's councils if direct elections were to be introduced. This problem was carefully considered by the policy makers involved. With regard to the Legislative Council, the risk was almost negligible since all of the proposed changes to this Council included built-in safeguards, such as restriction of the franchise to British subjects and the avoidance of direct elections. As far as Young's municipal council was concerned, the danger was (at least in principle) not so remote. In addition to providing direct elections based on a franchise which included non-British Chinese, the Young Plan also reserved two seats for labour leaders — representing both the left and right wings of the labour movement — thereby making the council potentially more vulnerable to being used as an arena for the debate of Chinese politics. Young planned to prevent such an event from happening by restricting the council to discussion of only municipal affairs. Although this could be done successfully in the early stages, the council could not be barred from debating issues with broader implications if it later developed into an alternative to the colonial government. However, this step would presumably not be taken unless policy makers at that later stage were satisfied that the elected municipal councillors would act responsibly in this matter. Since it was the policy of both the Kuomintang and the Communists not to destabilize Hong Kong, it is reasonably to believe that even if the Young Plan, not to mention the Unofficials' alternative or Grantham's 1950 proposals, had been implemented, Chinese politics would probably not have been brought into Hong Kong's councils.

Although there is no indication that the Kuomintang or the Communists intended to subvert the colony, the British officials were generally justified in taking the precautions they deemed necessary at that time. However, the so-called 'danger' of 1952, referred to by Grantham and the Unofficials, was a completely different matter. It consisted of a fairly remote risk of embarrassment to the several appointed ethnic Chinese unofficial members of the Legislative Council, and presumably also to the government, if some of those members were to take up anti-British rhetoric. It was not a real threat to the stability or well-being of Hong Kong and, on balance, the officials concerned were over-reacting.

Although neither the generally unstable situation in the Far East nor the potentially explosive circumstances in Hong Kong was the immediate cause of the abandonment of all major re-forms in 1952, they helped to create an atmosphere in which the abrupt turn-round of 1952 was not only possible but also non-controversial.

A Lost Chance?

Given the existing circumstances, could the two sets of proposals which had received the British government's approval — Gran-tham's 1950 proposals and the Young Plan — have succeeded in the ways their sponsors planned had they been implemented? Although it is not possible to say conclusively whether Gran-tham's 1950 proposals (or, for that matter, the Young Plan) could have worked successfully, two points are clear. First, policy mak-ers in Hong Kong and London thought that the 1950 scheme could have worked as planned. When the Foreign Office vetoed its implementation in early 1951, it did so purely for foreign policy reasons. Even when Grantham requested Lyttelton and the Colonial Office to back-pedal in the summer of 1952, he did not suggest that the scheme would have failed. Indeed, with the exception of Sidebotham, who preferred to let sleeping dogs lie, no senior British official dissented from the view that Lyttelton should recommend the scheme to the British Cabinet in the first half of 1952.

Secondly, the proposals were so meticulously worked out that there was little chance of failure. Since the right to nominate unofficial members to the Legislative Council was restricted to four local organizations — two of which, the Unofficial Justices of the Peace and the Hong Kong General Chamber of Com-merce, had for decades nominated two members in a very re-sponsible manner, while the third organization, the Hong Kong Chinese General Chamber of Commerce, consisted of Chinese capitalists who had no reason to rock the boat, and the fourth nominating body, the Urban Council, was dominated by the government-appointed Chairman (who was at that time usually a Cadet) — it was hardly likely that any of these organizations would have boycotted or sabotaged the scheme had it been implemented. There is also little chance that the associated Urban Council changes would have had a significantly higher possibility of failure since no fundamental change was proposed. In short, there is little ground for assuming that the 1950 scheme could not have worked successfully.

With regard to the Young Plan, some policy makers, most notably Grantham and officials in the Colonial Office, did have doubts as to whether the Plan was still appropriate for Hong Kong after 1948. Their main concern was that the Chinese Communists might infiltrate the proposed municipal council if it were to be introduced. Whilst the Colonial Office officials took this threat very seriously, it is unclear as to whether Grantham shared this view. In the same secret dispatch in which he pointed out this danger, Grantham also emphasized that he did not oppose the Young Plan as such; he objected only to its implementation before the duties and source of income of the proposed municipal council had been satisfactorily settled.[40] If Grantham had seriously expected the Young Plan to fail in the changed circumstances, he would have simply opposed it, but he did not. Grantham rejected the Young Plan because he thought that it represented the wrong approach, rather than because he thought it could not have worked.

An important consideration in whether the Plan could have been successfully implemented is timing. Had it been implemented in early 1948 — the time originally envisaged — it would probably have worked. At that time, the two Chinese political parties were preoccupied with their struggle inside China itself, and there was undoubtedly adequate local support to enable the proposed municipal council to function properly. Given the conflicting requirements of granting the local inhabitants (not just the British subjects) greater local self-government and preventing the introduction of Chinese politics into the various councils in Hong Kong, the Young Plan provided probably the most appropriate solution in the circumstances. Whether the two Chinese political parties would have attempted to infiltrate the municipal council after 1949 had it been established in 1948 or 1949 is, of course, a different question. Nevertheless, as indicated earlier, it is very unlikely that the Communists (or, for that matter, the Kuomintang) would have attempted to do so during this period.[41] On balance, therefore, it would appear that the Young Plan could have worked successfully during the period under study.

8 Conclusion

THE attempts at constitutional reform in Hong Kong during the immediate post-war years were first made jointly by Colonial Office officials and Hong Kong government officials then in exile in London because they thought them necessary as a result of the Pacific War. After the restoration of civil government in the colony, the task was entrusted to the Governor of Hong Kong, as was the usual practice. Although both the British government and the Unofficials played important roles in changing the various reform proposals, in the final analysis it was the two Governors who played the pivotal role. The turning-point was when Grantham succeeded Young as Governor in the summer of 1947. Since they had fundamentally different views on the matter, Grantham departed from Young's approach and eventually decided to abandon all attempts at major constitutional reform.

Whilst the deterioration in the socio-economic and political situation in the Far East and, to a lesser extent, in Hong Kong did not in itself cause Grantham and the other policy makers to decide against proceeding with any major reform, it was without doubt a very important consideration. The manner in which the articulate public in Hong Kong — who had hitherto played only a peripheral role — greeted this volte-face would indicate that at least the majority of these pressure groups were aware that the existing circumstances were not particularly conducive to political changes. Grantham finally decided against any major reform for four reasons. First, he did not think it was a good thing for Hong Kong. Secondly, the Unofficials had unanimously advised him not to proceed with the reforms. Thirdly, he thought that he could carry the day with the British government. Finally, he knew that there would not be strong public agitation against such a decision. All things considered, there is reason to think that had the proposals approved by the British Cabinet in May 1952 been implemented, they could have worked. Nevertheless, Grantham was justified in recommending the abandonment of all major reforms in the circumstances.

The ways in which these attempts at reform in Hong Kong were handled indicate that the Colonial Office recognized that

Hong Kong was a very special colony. There were four elements which together made Hong Kong unique. First and foremost, over ninety per cent of its total area was leased from China for a limited period only. Secondly, the Chinese wanted the return not only of the leased part, but also of the whole territory. Thirdly, both Chinese political parties which constituted the government of China at various times during the period maintained a strong presence in the colony. Finally, the main reason for the British presence in the colony was simply to facilitate and promote trading and other relations with China. Hong Kong was functionally just another treaty port along the China coast which happened to be (unlike Shanghai, for example) under exclusive British jurisdiction. In other words, with the exception of matters which were of purely local concern, almost all policy matters in relation to Hong Kong were bound up with considerations of Anglo-Chinese relations and, particularly from 1949 onwards, of Anglo-American relations. Britain and the United States had fundamentally different policies towards the new Communist government in China, which was not at that time unwilling to make friends with Britain, but was almost irrevocably hostile to the United States. This situation induced Grantham to think that Hong Kong should have been placed under Foreign Office control, rather than under the Colonial Office, but that it should be run by staff seconded from the Colonial Service.[1] The policy makers responsible for planning constitutional reform in Hong Kong had to address themselves to these considerations in addition to the usual problems associated with political development in the smaller colonial territories which, in the eyes of British officials, could not achieve dominion status and which had an overwhelmingly non-British population. Acceptance of the special character of Hong Kong on the part of the Colonial Office largely accounted for its willingness to rely to a very considerable extent on the advice of the Governors, to reject the Smaller Territories Committee's recommendations, and to desist from pressing Hong Kong to match the constitutional development then taking place in Singapore, which also had a predominantly Chinese population.

It must also be emphasized, however, that for a short time immediately after the war, the Colonial Office was confused about its policy for constitutional development in Hong Kong. Whilst dominion status was out of the question, there was a tendency to carry out in Hong Kong the same policy applied elsewhere. The 1 May 1946 statement that the residents of Hong Kong, like other peoples in the British Empire, should be given greater local self-government was unmistakably drafted with such an assumption. This situation was gradually rectified, however, particularly after Grantham became Governor. By October 1948, when Prime Minister Attlee complained that the Colonial Office, amongst others, had too readily assumed that the Westminster model of government could be reproduced in miniature in the smallest colonies, the Colonial Office had already accepted that Hong Kong was a special case. This acceptance was marked in the first instance by its approval in principle of Young's unorthodox proposals.[2]

The most important implication of the British government's 1952 decision that major reform was 'inopportune' for Hong Kong was that it marked the end of any attempt to establish in Hong Kong any system of government reminiscent of the Westminster model, for at least thirty years. Between 1952 and 1981, both the British and Hong Kong governments apparently ruled out any possibility of developing the Hong Kong constitution along the lines followed by most other British colonies.[3] The two governments decided that Hong Kong had to seek its own path in further political development, if and when that became desirable and necessary. As one example, when Sir David Trench (Governor, 1964–71) appointed a working party in 1966 to examine and recommend possible reforms, the scope of the inquiry was limited to 'that part of the general machinery of government which provides services or otherwise exercises responsibility designed to meet the collective needs of the residents of individual localities, as distinct from the needs of the population of the Colony as a whole'.[4] And again, when Sir Murray (now Lord) MacLehose reconstituted the Legislative Council during his governorship (1971–82), he was presumably still anxious to avoid committing the government to the more usual course of evolution in British colonies. MacLehose attempted to make the Legislative Council more representative of the population as a whole by appointing significantly more people with a non-business background to the Council. However, he steered well clear of introducing direct elections to the Council or expanding the practice of nominating, as a matter of course, representatives of selected public bodies such as the Hong Kong General Chamber of Commerce and the Unofficial Justices of the Peace to the Council. Indeed, he put an end to the practice of indirect election in order to avoid giving these two organizations special advantages. Mac-Lehose also made further changes by introducing an Unofficial majority of three into the Executive Council and by creating a *de facto* Unofficial majority of three in the Legislative Council. In addition, he pioneered, amongst other things, a district administration scheme in 1981. Nevertheless, he and the government avoided as far as possible any public discussion about constitutional reform, including the importation of the Westminster model and the revival of various proposals examined between 1945 and 1952. Up to the end of MacLehose's governorship, the Hong Kong government's annual year-books continued to state unequivocally that 'the British Government's policy towards Hong Kong is that there shall be no fundamental constitutional changes for which there is, in any event, little or no popular pressure'.[5]

Notes

Notes to Chapter 1

1. G. B. Endacott, *Government and People in Hong Kong: 1841–1962* (Hong Kong, Hong Kong University Press, 1964), pp. 22–5. Endacott's book is the best published source of information for constitutional developments in pre-war Hong Kong.
2. Endacott, *Government and People*, pp. 43–5 and 74–7.
3. Although Ng Choy, a British-trained Chinese barrister, had been nominated to the Legislative Council since 1880, Ng only held his seat on a temporary basis, in the absence of Gibb.
4. Endacott, *Government and People*, pp. 109–25.
5. Endacott, *Government and People*, pp. 135–45.
6. Endacott, *Government and People*, p. 14.
7. Quoted in Endacott, *Government and People*, p. 76.
8. Endacott, *Government and People*, pp. 126–62. See, also, L.A. Mills, *British Rule in Eastern Asia* (London, Humphrey Milford, 1942), pp. 373–422.
9. Quoted in Mills, *British Rule*, p. 391.
10. Endacott, *Government and People*, pp. 183–4.
11. Endacott, *Government and People*, p. 186.
12. Endacott, *Government and People*, pp. 196–8.
13. Endacott, *Government and People*, p. 196.
14. H. T. Cheung and C. K. Lo, *Political Reform in Hong Kong — The Past and the Future* (Hong Kong, Genius Publishing Company, 1984), p. 15.
15. Cheung and Lo, *Political Reform*, p. 19.
16. Cheung and Lo, *Political Reform*, p. 20.
17. References for these files are HKRS 41 D&S 1/2466, HKRS 41 D&S 1/4013, and HKRS 41 D&S 1/5902 respectively.

Notes to Chapter 2

1. K. E. Robinson, *The Dilemmas of Trusteeship* (London, Oxford University Press, 1965), p. 89.
2. J. M. Lee and M. Petter, *The Colonial Office, War and Development Policy* (London, Institute of Commonwealth Studies, 1982), pp. 162–3; and R. D. Pearce, *The Turning Point in Africa* (London, Frank Cass, 1982), pp. 35–6.
3. Command Paper 7167 (1947), p. 15; and Lee and Petter, *The Colonial Office*, p. 121.
4. FO 371/31804, F4320/4320/61, report from Ashley-Clarke to Eden, 11 June 1942.
5. FO 371/54052, F6208/2129/g.6, British Foreign Policy in the Far East (draft memorandum), 16 April 1946; and Command Paper 7709 (1949), p. 51.
6. CO 825/42/55104/2, various minutes, January–February 1942.
7. See FO 371/35905.
8. See FO 371/31662; FO 371/31665; CO 129/58823–4; and Chan Lau Kit-ching, 'The Hong Kong Question during the Pacific War (1941–1945)', *JICH*, Vol.2, 1973, pp. 56–74.
9. CO 825/42/55104/2, minute of Monson, 30 June 1943. For a general discussion of the American view during the war, see Chan Kit-ching, 'The United States and the Question of Hong Kong, 1941–1945, *JHKRAS*, Vol. 19, 1979, pp. 1–20.

10. CO 825/35/55104, minutes of Gent, 29 June and 1 July 1942, Sir G. Gater, 10 July 1942, and Lord Cranborne, 14 July 1942.

11. CO 825/55104/2, minutes and memoranda, July-November 1943; and FO 371/35824, F4541/G, minute of Ashley-Clarke, 3 September 1943.

12. See CO 129/592/8.

13. CO 825/35/55104/7; HKRS 211 D&S 2/8; F. S. V. Donnison, *British Military Administration in the Far East: 1943-1946* (London, HMSO, 1956), pp. 139 and 149; and MacDougall interview.

14. CO 129/591/9, letter from Lloyd to Sir Mark Young, 13 April 1946.

15. CO 129/592/8, minute of Ruston on the meeting, 1 May 1945.

16. HKRS 211 D&S 2/6; MacDougall interview; and FO 371/53630, F706/113/10.

17. CO 537/1650, minute of Ruston, 31 May 1945.

18. CO 537/1650, memorandum of HKPU, 26 June 1945.

19. CO 537/1650, memorandum of HKPU, 26 June 1945.

20. CO 537/1650, memorandum of Jones, and note of a meeting, 1 August 1945.

21. CO 537/1650, minutes of Ruston, 1 August 1945 and 18 January 1946.

22. Gimson had served six years in Ceylon, politically the most advanced of all British Asian colonies, before taking up his Hong Kong assignment on 7 December 1941, the day before the Japanese invaded Hong Kong. See also Chapter 7, pp. 190-2, for a comparison of the views of Gimson and his colleagues on the question of constitutional development in Hong Kong.

23. Mss Ind. Ocn. s. 222 (Gimson Papers) (Rhodes House Library, Oxford); and CO 537/1650, letter from Gimson to Ruston.

24. Mss 940.547252 G49 (Gimson Papers: Internment of Europeans at Stanley) (Hung On-to Memorial Library, Hong Kong); Mss 940.547252 G49i (Gimson Papers: Memories of Internment) (Hung On-to Memorial Library, Hong Kong); and G.B. Endacott and A. Birch, *Hong Kong Eclipse* (Hong Kong, Oxford University Press, 1978), pp. 344-65.

25. FO 371/46252-4; and correspondences, 16-27 August 1945, *FRUS 1945*, Vol. 7, pp. 500-13.

26. CO 537/1650, minute of Gent, 21 September 1945.

27. CO 537/1650, minute of Caine, 27 September 1945.

28. CO 129/592/8.

29. CO 825/35/55104, minutes of Gent, June and July 1942.

30. FO 371/53632, F3237/113/9.10, memorandum of Kitson, 28 February 1946. See also FO 371/31662 and FO 371/31665.

31. See FO 371/46259, F11795/1147/10 and F11807/1147/10; and FO 371/53630-4.

32. CO 537/1650, various minutes, January and February 1946.

33. CO 537/1650 and FO 371/53634, F6982/113/10, minute of Lloyd, 13 April 1946.

34. CO 129/591/16, telegrams from Foreign Office to Chungking, August 1945.

35. Mss Ind. Ocn. s. 222, ff. 169-80.

36. *The British Military Administration, Hong Kong, August 1945 to April 1946* (A Report to the Chief of Staff Committee); CO 129/591/18; CO 129/591/20; CO 129/592/6; CO 129/594/6; CO 129/595/9; FO 371/53634, F5435/113/10, minute of Kitson, 22 March 1946; Sir Cecil Harcourt, 'The Military Administration of Hong Kong', *JRCAS*, Vol. 34, 1947, pp. 7-18; private papers of G. A. C. Herklots (courtesy of Dr Herklots); MacDougall, Herklots and Cater interviews; and Endacott and Birch, *Hong Kong Eclipse*, pp. 261-78.

37. For example, see *WKYP* editorial, 12 May 1946.

38. For the former incident, see CO 129/592/6, Weekly Intelligence Summary, No.3; for the latter, see FO 371/53639, F16735/113/10.

39. CO 129/594/6, newspaper cuttings; CO 129/592/6, Weekly Intelligence Summaries; FO 371/53634, F5435/113/10, minute of Kitson, 22 March 1946; and FO 371/53638, F15424/113/G.10, letter from Sloss to Sir S. Cripps, 11 October 1946.

40. FO 371/46259, F11807/1147/10, notes by Keswick (on the future of Hong Kong and Secretary for Chinese Affairs).

41. See FO 371/46252–7, various minutes and correspondences, August–October 1945.

42. CO 129/592/6, Chief of Intelligence Report, 1 October 1945.

43. FO 371/46257, F8592/1147/10, note by Wallinger; and CO 129/592/6, Weekly Intelligence Summary, No. 7.

44. FO 371/53638, F15424/113/g.10; and HKRS 169 D&S 2/162.

45. HKRS 169 D&S 2/161; HKRS 169 D&S 2/162; CO 129/595/9, quarterly review by the CCAO, 15 December 1945; MacDougall interview; and Lin Yu-lan, *Hsiang-chiang shih-hua* (*A History of Hong Kong*) Hong Kong, Shanghai Book Store, 1980), pp.189–90.

46. CO 129/594/6, letter from CCAO to C-in-C, Hong Kong, 4 October 1945; and FO 371/46257, F8592/1147/10, note of Wallinger.

47. See E. Ride, *British Army Aid Group* (Hong Kong, Oxford University Press, 1981).

48. CO 129/595/9, report on the New Territories, September–November 1945; CO 129/592/6, Weekly Intelligence Summaries, Nos. 3 and 8; and FO 371/46257, F8592/1147/10, note of Wallinger.

49. *Hsin-wen-yen-chiu tsu-liao tsung-ti-shih-erh-chi* (*Newspaper Research Materials*) (Peking, News Research Institute of the Chinese Academy for the Social Sciences, 1982), Vol. 12, pp. 4–53 (hereafter cited as *NRM*).

50. HKRS 170 D&S 1/61.

51. HKRS 211 D&S 2/11; and CO 129/594/9, minute of Ruston, 12 March 1945.

52. Just prior to the outbreak of the war, the total population of the colony was estimated at 1.65 million, albeit including many Chinese refugees who had fled from the undeclared Sino-Japanese War since 1937.

53. Harcourt, 'The Military Administration of Hong Kong', *JRCAS*, Vol. 34, 1947, p. 15.

Notes to Chapter 3

1. CO 129/595/4, Governor's speech.

2. Hall was Colonial Secretary for only a short time, from July 1945 to October 1946, when he was succeeded by Arthur Creech-Jones.

3. CO 537/1651, dispatch from Young to Creech-Jones, 22 October 1946; CO 131/116, minute of Executive Council, 30 May 1946; and CO 537/1650, *HKSH*, 30 June 1946.

4. All major Hong Kong newspapers, 29 August 1945, e.g. *WKYP*.

5. CO 537/1651, dispatch from Young to Creech-Jones (No. 145), 22 October 1946.

6. CO 537/1651, dispatch from Young to Creech-Jones (No. 145), 22 October 1946.

7. CO 537/1651, dispatch from Young to Creech-Jones (No. 145), 22 October 1946.

8. See pp. 45–6.

9. CO 537/1651, dispatch from Young to Creech-Jones (No. 145), 22 October 1946.

10. CO 537/1651, confidential dispatch from Young to Creech-Jones, 22 October 1946. For a discussion on public responses, see pp. 59–62.

11. CO 537/1651, confidential dispatch from Young to Creech-Jones, 22 October 1946.

12. CO 537/1651, confidential dispatch from Young to Creech-Jones, 22 October 1946. See also CO 131/116, minute of Executive Council, 2 August 1945.

13. CO 537/1651, confidential dispatch from Young to Creech-Jones, 22 October 1946.

14. CO 537/2188, Young to Creech-Jones, 7 February 1947.

15. CO 537/2188, Young to Creech-Jones, 7 February 1947.

16. CO 537/1651, minute of Ruston, 18 November 1946.

17. CO 537/1651, minute of Mayle, 6 December 1946.

18. See also pp. 42-3.

19. CO 537/1651, minute of Caine, 16 December 1946, and minute of a meeting, 18 December 1946. For the more usual way of introducing constitutional advance in British colonies, see A.F. Madden, 'How Colonies Grow into Dominions', *United Empire*, Vol. 41, No. 3, 1950, pp. 159-64.

20. CO 537/1651, minute of Lloyd, 12 December 1946.

21. CO 537/1651, minutes, November–December 1946.

22. CO 537/1651, minute of a meeting, 18 December 1946.

23. CO 537/1651, minutes of Creech-Jones, 13 and 18 January 1947.

24. D. Goldsworthy, *Colonial Issues in British Politics: 1945-1961* (Oxford, Oxford University Press, 1971), p. 19.

25. CO 537/1651, minute of Creech-Jones, 18 January 1947.

26. CO 537/1651, various minutes, January 1947, and Creech-Jones to Young, 24 January 1947.

27. CO 537/2188, Young to Creech-Jones, 7 February 1947.

28. CO 537/2188, various minutes, 14–18 February 1947.

29. FO 371/63388, F5527/376/10, minute of Scott, 24 April 1947.

30. FO 371/63388, F5527/376/10, minute of Kitson, 27 April 1947, and his letter to Mayle, 28 April 1947.

31. CO 537/2188, minute of Mayle, 3 March 1947.

32. CO 537/2188, various minutes of Colonial Office and telegraphic exchanges between Creech-Jones and Young, 21 February–25 April 1947.

33. CO 537/2188, various minutes of Colonial Office and telegraphic exchanges between Creech-Jones and Young, 2 April–30 June 1947.

34. CO 537/2188, various minutes, July 1947.

35. CAPMC 1947, minutes and circulars of Executive Committee, 28 February–21 July 1947. For the views of the China Association, see pp. 58-9.

36. CO 537/1656, Status of Hong Kong (and other papers).

37. FO 371/53632, F3237/113/G.10; FO 371/53633, F5107/113/G.10; FO 371/53635, F10572/113/G.10, Future of Hong Kong (third draft of a memorandum), 18 July 1946; and FO 371/53637, F12400/113/G.10, secret letter from Lloyd to Dening, 22 August 1946.

38. CO 537/1651, minute of Creech-Jones, 18 January 1947.

39. Ambassador in China to Secretary of State, 4 March 1947, *FRUS 1947*, Vol. 7, p. 55.

40. Caine interview.

41. CO 537/1651, dispatch from Creech-Jones to MacDougall, 3 July 1947.

42. CO 129/594/4, dispatch from Young to Hall, 8 May 1946; CO 129/594/9, minute of Ruston, 16 May 1946; *Hong Kong Hansard 1946* (Hong Kong, 1946), minutes, 1 and 16 May 1946; CO 131/116, minutes, 11 May 1946; HKRS 211 D&S 2/11; and Endacott and Birch, *Hong Kong Eclipse*, pp. 279-82.

43. CO 129/594/9, minute of Ruston, 16 May 1946; and *Hong Kong Hansard 1946*, p. 63. For a general study of localization, see D. Podmore, 'Localization in the Hong Kong Government Service, 1948-1968', *JCPS*, Vol. 9, 1971, pp. 36-51.

44. Colonial Office, *Annual Report on Hong Kong for the Year 1946* (London, 1948) p. 2 (hereafter cited as '*Annual Report 1946*'); and MacDougall, Sedgwick, Herklots, and Cater interviews. See also G. C. Hamilton, *Government Departments in Hong Kong: 1841–1969* (Hong Kong, Government Printer, 1969). For an introduction to the Hong Kong Cadet Service before the war, see H. J. Lethbridge, *Hong Kong: Stability and Change* (Hong Kong, Oxford University Press, 1978), pp. 31–51.

45. CO 129/611/2, dispatch from Young to Creech-Jones, 15 November 1946. See also various minutes in the same file.

46. *Hong Kong Hansard 1946*, pp. 109–18; and *Hong Kong Hansard 1947*, pp. 68–83, 118–28, and 133–56. See also CO 129/595/3; HKRS 41 D&S 1/1007, letter from Lo to MacDougall, 2 May 1947; and *FEER*, Vol. 1, 1946, p. 1, and Vol. 2, 1947, pp. 141–3.

47. *Annual Report 1946*, p. 26; Census and Statistics Department, *Hong Kong Statistics 1947–1967* (Hong Kong, Census and Statistics Department, 1969), pp. 88 and 209 (hereafter cited as '*Hong Kong Statistics*'); and *FEER*, Vol. 1, 1946, No. 11, p. 1, and Vol. 2, 1947, pp. 55 and 106.

48. *Hong Kong Hansard 1946*, pp. 21 and 48–9.

49. Wah Kiu Yat Pao, *Hsiang-chiang nien-chien* (*Hong Kong Year Book*), Vol. 1, 1948 (Hong Kong, Wah Kiu Yat Pao, 1948), Section 1, pp. 6–7, and section 6, pp. 8–14 (hereafter cited as '*Wah Kiu Year Book*'); and *Hong Kong Statistics*, p. 144.

50. For an example, see *Diary of Mr John Swire's Post-war Trip to China and Australia: 25 January to 1 July 1946* (courtesy of Sir Adrian Swire, John Swire and Sons Ltd, London).

51. CO 537/1651, confidential memorandum of Megarry on Kuomintang activities, 27 November 1946.

52. '*National Times*' is the English title for the newspaper, even though, literally translated, the Chinese title of the newspaper means 'National People's Daily'.

53. CO 537/2193, savingrams from Young to Hall, May–July 1946, and from Young to Creech-Jones, November–December 1946; CO 131/116, minutes of Executive Council, 8 June and 18 December 1946; and FO 371/53635, F9445/113/10. See also CO 537/1658.

54. Chung-kuo kuo-min-tang chu-chiang-ao tsung-chih-pu, *Tang-yuan tung-hsin* (Hong Kong Branch of the Kuomintang, *Party Bulletin*), No. 7, 21 June 1947, p. 6; and No. 8, 1 July 1947, p. 8 (hereafter cited as '*KMT Bulletin*').

55. G.W. Catron, *China and Hong Kong, 1945–1967* (Harvard University Ph. D. thesis, 1971), pp. 46–9; *Wah Kiu Year Book 1948*, Section 1, p. 7; *FEER*, Vol. 2, 1947, pp. 289–90, 333–4, and 377–8.

56. *KMT Bulletin*, Nos. 9 and 10 combined issue, 21 July 1947.

57. Speech by Li Ta-chao, *KMT Bulletin*, Nos. 9 and 10 combined issue, 21 July 1947, pp. 26–9.

58. Chung-kuo kuo-min-tang chu-chiang-ao tsung-chih-pu, *Shuang-shih te-kan* (Hong Kong and Macau Branch of the Kuomintang, *Double-Tenth Special Publication*) (Hong Kong, 1946).

59. CO 129/592/8, extract from address of President Chiang to the Supreme National Defence Council, 25 August 1945.

60. FO 371/63388, F6386/376/10, Sir R. Stevenson to Bevin, 8 May 1947.

61. *HSWP*, 25 September 1946, and *KMJP*, 25 July 1947, in *Chinese Press Review 1947* (prepared by American Consulate-General in Hong Kong, copy in University Services Centre, Hong Kong) (hereafter cited as '*KMJP*, 25 July 1947, in *CPR 1947*'); *China Mail*, 20 November 1946, in Rhodes House Library, Mss Brit. Emp. s. 365 (Fabian Colonial Bureau Papers), Box 152, File 1.

62. FO 371/63391, F13158/376/10, secret dispatch (and enclosure) from Grantham to Stevenson, 27 August 1947.

63. CO 537/3713, secret dispatch from Grantham to Creech-Jones, 28 May 1948.

64. FO 371/63388, F6487/376/10, top-secret dispatch from Young to Creech-Jones, 17 April 1947.

65. Catron, *China and Hong Kong*, p. 72; Union Research Institute, *Who's Who in Communist China* (Hong Kong, Union Research Institute, 1965), pp. 117, 178–9, and 373; *NRM*, Vol. 12, p. 28; and memorandum from Consul-General, Hong Kong to General Wedemeyer, 13 August 1947, *FRUS*, Vol. 7, 1947, pp. 720–1.

66. FO 371/63387, F3337/376/10, minute of conversation with MacDougall, 8 February 1947; CO 537/2197, letter from Galsworthy to Scott, 8 August 1947; and *Report of the Commissioner of Police for the Year 1947–1948* (Hong Kong, Government Printer, 1948).

67. *HSP* editorial, 25 July 1947.

68. *SCMP* letters from Simple Simon, 4 May 1946, and Peter Patter, 13 May 1946.

69. *SCMP* editorials, 8 and 20 September 1946; and *FEER*, Vol. 1, 1946, No. 1, p. 1 and No. 11, p. 1; and Vol. 2, 1947, pp. 117, 141, and 193–4.

70. HKGCC, *Report for the Period 1941–1946* (Hong Kong, Ye Olde Printerie Limited, 1947), pp. 89–96.

71. CAPMC 1947, circulars of Executive Committee, 3 February, 2 April, and 21 July 1947, letter to the Colonial Office, 28 April 1947, and letter from Keswick to Creech-Jones, 10 July 1947.

72. The third one, Watson, was a nominee of the Unofficial Justices of the Peace.

73. *Hong Kong Hansard 1947*, p. 139.

74. *Hong Kong Hansard 1947*, p. 72; and *Hong Kong Hansard 1946*, p. 118.

75. *Hong Kong Hansard 1947*, p. 143.

76. *China Mail*, 20 November 1946, Mss Brit. Emp. s. 365, Box 152, File 1.

77. See also Chapters 5 and 7, particularly pp. 91–9 and 196–9.

78. *WKYP* editorial, 1 May 1946; *KSJP* editorial, 1 May 1946; and *HSP* editorial, 1 May 1946. For a good account of the view of the Hong Kong Chinese of the British officials, see A. Coates, *Myself a Mandarin* (Hong Kong, Heinemann Asia, 1975; reprinted by Oxford University Press, Hong Kong, 1987).

79. *WKYP* editorials, 12 May and 13 June 1946.

80. *HSWP*, 28 September 1946; and CO 537/1650, *HKSH*, 30 June 1946.

81. *HSWP*, 21, 28, and 29 September 1946, and 1, 2, and 26 October 1946; and *Chinese Engineers' Institute Bi-monthly* (in Chinese, but English subtitle is original), Vol. 1, No. 2, 1946, pp. 1–2.

82. For a good introduction to Chinese political culture, see R.H. Solomon, *Mao's Revolution and the Chinese Political Culture* (Berkeley, University of California Press, 1971), pp. 28–153, particularly pp. 49–60 and 144–50.

83. 'Basic Concepts and Comments on Self-government Proposals', Mss Brit. Emp. s. 365, Box 152, File 1.

Notes to Chapter 4

1. Governor's arrival speech, *SCMP*, 26 July 1947.

2. Governor's speech (31 July), *Hong Kong Hansard 1947*, p. 256.

3. CO 129/609/2, minute of Galsworthy, 1 August 1947.

4. A. W. G. H. Grantham, 'Hong Kong', *JRCAS*, Vol. 46, 1959, p. 121. For an in-depth analysis of Grantham's views, see Chapter 7, pp. 186–90.

5. Because of the Hong Kong government's embargo on files relating to the reforms, it is not possible to ascertain whether Grantham had discussed the matter with his official advisers or not before the middle of August. According to the records of the Legislative and Executive Councils, he had not officially discussed the matter with his unofficial advisers.

6. CO 882/31, secret dispatch from Governor to Secretary of State, 25 August 1949; CO 537/2188, Grantham to Creech-Jones, 15 August 1947; and CO 131/117, minute of 14 May 1947. The draft bills are available in CO 129/609/3.

7. *Civil Service List for 1950–1* (Hong Kong, Government Printer, 1951), p. 426; CO 537/2189, letter from Grantham to Mayle, 17 November 1947; and CO 131/120, minute of Executive Council, 30 November 1948.

8. For the minutes of the Executive Council, July 1947–June 1948, see CO 131/118 and CO 131/119.

9. CO 537/2188, minute of Galsworthy, 12 September 1947; CO 129/609/2, minute of Galsworthy, 1 August 1947, and letter from Mayle to MacDougall, August 1947; and CO 537/2189, various minutes, October 1947, letter from Mayle to Grantham, 22 October 1947, and letter from Mayle to Pitblado (Treasury), 22 October 1947.

10. CO 537/3703, letter from Pitblado to Mayle, 18 March 1948.

11. CO 537/3703, dispatch from Creech-Jones to Grantham, 27 March 1948.

12. *Hong Kong Hansard 1948*, pp. 57–8.

13. *STJP*, 25 July 1947.

14. CO 537/2188, enclosure in savingram from Young to Creech-Jones, 29 April 1947.

15. For the 1952 Urban Council elections, see Chapter 6, pp. 158–61.

16. CO 537/3703, minute of Wallace, 9 November 1948, and Grantham to Creech-Jones, 1 December 1948.

17. *Civil Service List 1951*, p. 426.

18. CO 537/3707, minute of Wallace, 3 June 1948; Grantham's original letter (to Mayle, 17 November 1947) is in CO 537/2189.

19. CO 537/2189, confidential telegram from Grantham to Creech-Jones, 18 June 1948.

20. See pp. 73–4.

21. CO 537/3703, minutes of Wallace, 3 June and 9 November 1948.

22. CO 537/3703, dispatch from Grantham to Creech-Jones, 24 August 1948.

23. CO 537/3702, various minutes, June 1948.

24. CO 537/3327, Lamb (Minister in Nanking) to Bevin, 28 October and 18 November 1948; CO 537/2657, MI2 Intelligence Report, 3 November 1948; CO 537/3295, FE(O)C note, 10 December 1948; and CAB 129/31, CP 80(48)299, 9 December 1948. See also CO 537/3325, CO 537/3326, CO 537/3328, and CO 537/3329 for the British observations of events in China which might affect Hong Kong.

25. CO 537/3703, minutes of Wallace, 9 November 1948, and Sidebotham, 10 November 1948; CO 537/3726, note of Wallace, 30 March 1948; and CO 37/3327, minute of Paskin, 26 November 1948.

26. CO 131/120, minute of Executive Council, 30 November 1948; and CO 537/3703, Grantham to Creech-Jones, 1 December 1948.

27. FO 371/69583, F16417/154/10, minute of Bagley, 20 December 1948.

28. 'Presumably' because the 1949 Colonial Office file on the reforms is still closed.

29. CO 537/3703, all minutes, December 1948; and CO 882/31, confidential dispatch from Grantham to Creech-Jones, 25 August 1949.

30. *Civil Service List 1951*, p. 426.

31. CO 882/31, secret dispatch from Grantham to Creech-Jones, 25 August 1949.

32. *Civil Service List for 1950–1*, pp. 389 and 426.

33. See M. Wight, *The Development of the Legislative Council, 1606–1945* (London, Faber, 1946), p. 109.

34. See also Chapter 7, particularly pp. 187–90 and 203–8.

35. Editorials of *HSWP* and *HSP*, 25 July 1947, and *STJP* and *KSJP*, 26 July 1947.

36. Editorials of *HSWP*, 24 and 25 July 1947, *STJP*, 25 and 26 July 1947, and *KSJP*, 25 and 26 July 1947.

37. *SCMP* editorial, 25 July 1947; and *FEER*, Vol. 3, 1947, pp. 193–4 and 225–6.

38. Editorials of *HSWP*, 17 March 1948, *HSP*, 18 March 1948, *KSJP*, 19 March 1948, *STJP*, 20 March 1948, and *WKYP*, 24 March 1948.

39. *STJP* news, 21 March 1948.

40. CO 537/3702, minute of Mayle, 19 March 1948; for Listowel's visit, see HKRS 41 D&S 1/4545.

41. *HKT*, 17 August 1948, in CAPMC 1948, Bulletin 28.

42. See Chapter 5, pp. 143–7.

43. Morse's speech, *SCMP*, 26 July 1947.

44. CO 537/3703, minute of Sidebotham, 10 November 1948.

45. CO 537/3702, letter from Grantham to Sidebotham, 19 May 1948.

46. Although constitutional reform might also mean that the British would withdraw in due course, such an eventuality had been specifically ruled out by Grantham, who had publicly declared, during an official visit to the Chinese government in 1947, that Hong Kong would not achieve responsible government and move towards independence, but would remain a colony after the reforms. FO 371/63391, F14181/376/10.

47. *Hong Kong Statistics*, pp. 88, 97–8, 209, and 211–12; HKGCC, *Report for 1947*, p. 9; HKGCC, *Report for 1948*, p. 9. The value of trade in constant price is not available.

48. *Hong Kong Statistics*, pp. 48–9; *Annual Report 1948*, p. 62; and CO 537/3326, Review of Chinese Affairs, May 1948.

49. Sir A. Grantham, *Via Ports* (Hong Kong, Hong Kong University Press, 1965), pp. 103–4; and Sedgwick interview (2 February 1983). The exact value of the entrepôt trade cannot be ascertained because the value of re-export was listed as part of export in those days. See Table 6.1 in *Hong Kong Statistics*, p. 88.

50. The average price index for both 1947 and 1948 is lower than 100 because when the index was first prepared in March 1947, the price level was still artificially high as a result of the post-war shortages of various essential goods such as rice and firewood.

51. *Hong Kong Statistics*, p. 144; *Hong Kong Annual Report of the Commissioner of Labour (1948–1949)* (Hong Kong, 1949), pp. 29–30 and 34 (hereafter cited as '*Labour Report 1948–1949*'); *Annual Report on Hong Kong for the Year 1948* (Hong Kong, HMSO, 1949), pp. 22–3.

52. Financial Secretary's budget speech, *Hong Kong Hansard 1948*, p. 64.

53. *FEER*, Vol. 5, 1948, pp. 643–4; and *Annual Report 1949*, p. 85.

54. CO 537/3717, dispatch from Grantham to Creech-Jones, 25 May 1948.

55. Convention of Peking, in P. Wesley-Smith, *Unequal Treaty, 1898–1997* (Hong Kong, Oxford University Press, 1983), p. 191.

56. CO 537/1657, notes of Gent, 22 July 1945, and of Foreign Office, 19 December 1946. See also Wesley-Smith, *Unequal Treaty*, pp. 57–82 and 123–6.

57. CO 537/1557, various telegrams and reports from Young to Creech-Jones, September 1946, and from Sir R. Stevenson (Ambassador in China) to Bevin, 24 September 1946.

58. Grantham, *Via Ports*, pp. 130–1.

59. FO 371/69578, F1393/154/10, note from Colonial Office, 19 January 1948; *HSWP*, 4 December 1947; Li Chin-wei, *Centenary History of Hong Kong* (in Chinese, but English subtitle is original) (Hong Kong, 1948), pp. 98–101; and Catron, *China and Hong Kong*, pp. 54–5.

60. FO 371/69578, F1393/154/10, note from Colonial Office, 19 January 1948; FO 371/69577, F630/154/10, Stevenson to Bevin, 13 January 1948; CO 537/3709, Consul-General's Report; all editorials and news on the Walled City, 8 December 1947–18 January 1948, in *CPR 1947 and 1948*; Grantham, *Via Ports*, pp. 131–2; Catron, *China and Hong Kong*, pp. 55–7; and various telegrams from Canton, Shanghai, and Nanking to the Secretary of State, 16, 19, 21, 22, 29, and 30 January 1948, *FRUS 1948*, Vol. 7, pp. 47–68.

61. All editorials on the Walled City, 8 December 1947 – 18 January 1948, in *CPR 1947 and 1948*.

62. Catron, *China and Hong Kong, pp. 57–8*; Hwa Shiang Pao, *I-chiu-shih-chi shou-cha (1947 Year Book)* (Hong Kong, Hwa Shiang Pao, 1948), pp. A24–5, A92, and C19; and M. Perleberg, *Who's Who in Modern China* (Hong Kong, Ye Olde Printerie Limited, 1954), pp. 171–2. For a good introduction to various factions in the Kuomintang, see H. M. Tien, *Government and Politics in Kuomintang China, 1927–1937* (Stanford, Stanford University Press, 1972), pp. 45–72.

63. FO 371/75806, F3662/10129/10g, dispatch from Grantham to Creech-Jones, 24 February 1949.

64. CAB 129/32, Appendix to CP (49)39, 4 March 1949; FO 371/69582B, F10787/154/10, savingram from Stevenson to Bevin, 21 July 1948; and *KMJP*, 12 July 1948, in *CPR 1948*.

65. FO 371/75780, F5279/1016/10, dispatch from Grantham to Creech-Jones, 2 April 1949; FO 371/75779, F1180/1016/10; CO 131/120, minutes, 16 and 23 November 1948; CO 131/121, minutes, 22 February 1949; and Catron, *China and Hong Kong*, pp. 93–7.

66. CO 537/3726, note of Wallace, 30 March 1948, and minute of Lord Listowel, 28 April 1948; FO 371/75779, F3690/1016/10g, dispatch from Grantham to Creech-Jones, 23 February 1949.

67. *HSP* editorials, 6, 13, and 18 January 1948, in *CPR 1948*.

68. FO 371/75779, F124/1016/10, confidential memorandum of Bough.

69. CAB 129/32, CP (49)39; and FO 371/75779, F3690/1016/10g, dispatch from Grantham to Creech-Jones, 23 February 1949.

70. CO 537/2188, memorandum of Hawkins, 10 June 1947.

71. CO 537/2188, memorandum of Hawkins, 10 June 1947.

72. *Labour Report 1947–1948*, pp. 15–16.

73. See *FEER*, Vol. 3, 1947, pp. 289–90, 333, 377, and 381; Vol. 4, 1948, pp. 509–10; Vol. 5, 1948, p. 47; and Vol. 6, 1949, pp. 102 and 169.

74. Chiang-chiu kung-tuan lien-ho-tsung-hui (Hong Kong Trade Union Council), *Hong Kong Labour* (in Chinese, but English subtitle is original), 1949, No. 1, p. 1.

75. Letters from Baker to Fabian Colonial Bureau, 19 December 1947 and 6 April 1948, in Mss Brit. Emp. s. 365, Box 152, Files 3 and 1 respectively.

76. *Labour Report 1948–1949*, p. 20.

77. *KMJP* editorial, 9 September 1948, in *CPR 1948*.

Notes to Chapter 5

1. *Hong Kong Hansard 1949*, pp. 62, 91, and 137–8; and CO 882/31, dispatch from Governor to Secretary of State, 25 August 1949.

2. *Hong Kong Hansard 1949*, p. 150.

3. CO 1023/41, minute of Paskin, 10 March 1952; see also Chapter 4, pp. 70–1.

4. *Hong Kong Hansard 1949*, pp. 151 and 188–90.

5. Lo's speech, *Hong Kong Hansard 1949*, pp. 188–96.

6. See pp. 143–8.

7. Grantham, *Via Ports*, p. 111.

8. Lo's speech, *Hong Kong Hansard 1949*, pp. 188–96.

9. Lo's speech, *Hong Kong Hansard 1949*, pp. 188–96.

10. See p. 101.

11. Lo's speech, *Hong Kong Hansard 1949*, pp. 188–96.

12. It is not entirely clear to whom Lo were referring: presumably, they were the Reform Club and the Chinese Reform Association and their supporters who asked for the direct election of all unofficial members in the second stage of their different (but also similar) two-stage Legislative Council reform proposals. See pp. 144–8.

13. Lo's speech, *Hong Kong Hansard 1949*, pp. 188–96.

14. Watson's speech, *Hong Kong Hansard 1949*, pp. 196–9.

15. Castro's speech, *Hong Kong Hansard 1949*, pp. 201–3.

16. E. Hambro, *The Problems of Chinese Refugees in Hong Kong* (Leydon, A. W. Sijthoff, 1955), p. 149, Table X.

17. Blaker's speech, *Hong Kong Hansard 1949*, p. 203.

18. Governor's speech, *Hong Kong Hansard 1949*, p. 137.

19. T. N. Chau's speech, *Hong Kong Hansard 1949*, p. 199.

20. S. N. Chau's speech, *Hong Kong Hansard 1949*, pp. 199–201.

21. Governor's speech, *Hong Kong Hansard 1949*, p. 204.

22. CO 882/31, dispatch from Governor to Secretary of State, 25 August 1949.

23. CO 882/31, secret dispatch from Governor to Secretary of State, 25 August 1949.

24. CO 882/31, secret dispatch from Governor to Secretary of State, 25 August 1949.

25. CO 882/31, secret dispatch from Governor to Secretary of State, 25 August 1949.

26. CO 882/31, secret dispatch from Governor to Secretary of State, 25 August 1949. It is noteworthy that the results of Urban Council elections, including those of the first post-war elections held in 1952, were generally not based on ethnic grounds. For the first post-war elections, see Chapter 6.

27. For details, see pp. 147–9.

28. CO 882/31, secret dispatch from Governor to Secretary of State, 25 August 1949.

29. CAB 128/13, CM 80(48)3, 13 December 1948; and DEFE 6/7, JP (48)124 (Final), 12 January 1949.

30. DEFE 5/13Pt1, annex to COS (49)12, 7 January 1949.

31. CAB 129/13, CM 80(49)3; and CAB 19/31, CP (48)299 and CP (49)39.

32. See CAB 128/15, CM 28(49)1; CAB 129/34, CP (49)93; FO 371/75887–97; and L. Earl, *Yangtze Incident* (London, London & Co., 1950).

33. Despite the British media's portrait of the escape of the frigate as a great anti-Communist victory, the incident was seen as a humiliation of the Royal Navy by the Far Eastern people who did not expect a British warship to be held hostage for over a hundred days before it could slip away under cover of darkness.

34. CAB 129/35, CP (49)118, 24 May 1949.

35. CAB 128/15, CM 30(48)4, 28 April 1949, CM 32(49)2, 5 May 1949, CM 33 (49)2, 9 May 1949, and CM 38(49)3, 26 May 1949; and DEFE 6/9, annex to JP (49)50 (Final).

36. 464 HC DEB 5s, cc. 1250–1; CAB 128/15, CM 38(49)3, 26 May 1949, and

CM 42(49)5, 23 June 1949; CAB 129/35, CP (49)134, 17 June 1949; and FO 371/75877, F10083/1192/10, top-secret and personal savingram from Creech-Jones to Grantham, 6 July 1949.

37. CAB 128/16, CM 54(49)2, 29 August 1949; CP (49)177, in Mss Brit. Emp. s. 332 (Creech-Jones Papers) (Rhodes House Library, Oxford), Box 57, File 1; and CAB 128/16, CM 54(49)2.

38. CO 537/6046, minutes of Sidebotham, 11 January 1950, Sir K. Roberts-Wray, 2 February 1950, Radford, 30 March 1950, and Paskin, 5 May 1950; CO 537/6491, minute of Paskin, 11 November 1949; 469 HC DEB 5s, cc. 208 and 381; and 470 HC DEB 5s cc. 1138 and 2641–2.

39. CO 537/6046, minute of Creech-Jones, 6 January 1950. For details about the Smaller Territories Committee, see pp. 119–24.

40. CO 537/6046, minute of Sidebotham, 11 January 1950; Grantham to Creech-Jones, 17 January 1950; and minute of Paskin, 19 January 1950.

41. CO 537/6046, all minutes, 28 February–end of April 1950.

42. CO 537/6046, minutes of Paskin and Jeffries, 5 May 1950. See also pp. 124–40 for details.

43. CO 537/6046, minutes of Cook and Griffiths, 17 May 1950, Jeffries, 19 May 1950, Sidebotham, 23 May 1950, and note of discussions in Paskin's room, 26 June 1950.

44. CO 537/6046, note of discussion with Nicoll, 30 January 1950, note of discussion in Paskin's room, 21 June 1950, and minute of Paskin, 24 June 1950; and CO 131/123, minute of Executive Council, 6 June 1950.

45. CO 537/6046, minutes of Sidebotham, 23 June 1950, and Paskin, 26 June 1950.

46. CO 537/6046, Griffiths to Grantham, 9 May 1950, and notes of meetings with Grantham, 21 and 26 June 1950.

47. CO 537/6046, notes of meetings in Griffiths' room, 30 June and 4 July 1950.

48. CO 537/6046, notes of meetings in Griffiths' room, 30 June and 4 July 1950.

49. This number, 139, is given in the Chinese Reform Association's letter, but records of 1949 suggest that it should be 141, including the three co-sponsors of the petition.

50. CO 537/6046, notes of meeting in Jeffries' room, 13 July 1950, and in Cook's room, 19 July 1950, and minutes of Paskin, 2 and 9 August 1950; and CO 537/6047, letter from Hong Kong Chinese Reform Association to Sir H. Poynton, 20 July 1950.

51. CO 537/6046, notes of meeting in Jeffries' room, 13 July 1950, and in Cook's room, 19 July 1950; letter from Grantham to Cook, 27 July 1950; and letter from Paskin to Grantham, 11 August 1950.

52. CO 537/6046, letters from Paskin to Nicoll, 14 August 1950, and Nicoll to Paskin, 6 September 1950; the relevant Executive Council minute is in CO 131/124.

53. CO 537/6046, note of a meeting, 20 September 1950, and letter from Sidebotham to Nicoll, 10 October 1950.

54. CO 537/6046, letters from Grantham to Paskin, 7 November 1950, and Paskin to Grantham, 1 December 1950; and CO 131/124, minute of Executive Council, 31 October 1950.

55. CO 537/6046, letters from Grantham to Paskin, 7 November 1950, and from Hall to Trench (Foreign Office), 1 December 1950.

56. CO 537/6046, minute of Paskin, 25 November 1950.

57. FO 371/83268, FC 10117/2, minute of Trench, 18 December 1950.

58. CO 537/6046, letter from Shattock to Sidebotham, 28 December 1950.

59. CO 537/6046, minutes of Hall, 3 January 1951, and letter from Paskin to Grantham, 5 January 1951.

60. CO 537/6074, all minutes, July 1950; FO 371/83296, FC1023/6/G and FC1023/8/G; DEFE 6/13, JP (50)82 (Revised Final), 5 July 1950; and CO 537/6305, all minutes, December 1950. For a countdown to the Chinese interference in the Korean War, see A. S. Whiting, *China Crosses the Yalu* (Stanford, Stanford University Press, 1960).

61. FO 371/92213, FC 10110/3, letter from Grantham to Paskin, February 1951.

62. CO 882/31, dispatch from Grantham to Creech-Jones, 25 August 1949.

63. Endacott, *Government and People*, pp. 198–9.

64. D. J. Morgan, *The Official History of Colonial Development* (London, Macmillan and Co., 1980), Vol. 5, pp. 35–7.

65. Morgan, *Colonial Development*, pp. 32–5; CO 537/4389, Creech-Jones' opening speech; CO 537/5396, brief for the Secretary of State when he met the Smaller Territories Committee; CAB 129/33, annex to CP (49)62 (Memorandum by Colonial Secretary), 10 March 1949.

66. CO 967/146, Interim Recommendations by Sir F. Rees, March 1951, and Report of the Committee, August 1951; CO 967/149, analysis of comments made in the Colonial Office and by Governors; and Morgan, *Colonial Development*, pp. 42–51.

67. CO 537/4398, STC (49), third minutes, 12 December 1949.

68. CO 537/5400, minute of sixth meeting, held 28–9 March 1950; CO 537/5396, note by Reid on Hong Kong, April 1950; and CO 537/5397, notes by Sir J. Maude on Hong Kong, April 1950, and by Harlow on Hong Kong, 12 July 1950.

69. CO 537/5397, note by Harlow on Hong Kong, 12 July 1950.

70. CO 537/5400, minute of ninth meeting, held 11–12 July 1950; and CO 967/146, confidential report to the Secretary of State (paragraph 103).

71. For an example, see CO 537/3703, minute of Wallace, 9 November 1948.

72. Grantham, *Via Ports*, pp. 137–8.

73. CO 537/3702, Grantham to Creech-Jones, 24 December 1948; and FO 371/75780, F6927/1016/10G, dispatch from Grantham to Creech-Jones, 3 May 1950.

74. CO 537/3729, minute of Paskin, 9 December 1948; CAB 129/35, CP (49)120, 23 May 1949, and CP (49)134, 17 June 1949; CO 537/4824, Grantham to Creech-Jones, 1 April 1949; FO 371/75783, F12541/1016/10, note by the Colonial Office, 12 August 1949; FO 371/83262, FC 10112/45; and *FEER*, Vol. VII, 1949, p. 324. For an outline of Hong Kong's policy in response to any Chinese Communist challenge, see CO 537/4847, memorandum of the Political Adviser enclosed in dispatch from Grantham to Creech-Jones, 17 October 1949.

75. CO 537/5240, minute of Paskin, 29 June 1950; FO 371/83260, FC 10112/25G, report on Chinese Communist activities ending 31 December 1949; and CO 537/4328, Colonial Political Intelligence Summary, March–April 1949.

76. CO 131/121, minute of Executive Council, 22 April 1949.

77. FO 371/75839, F6195/1061/10G, Grantham to Creech-Jones, 30 April 1949 (from MacDonald); 464 HC DEB 5s c. 1299, speech by Gammans, 5 May 1949; FO 371/75780, F4788/1016/10G, Grantham to Creech-Jones, 21 May 1949, and F6927/1016/10G, dispatch from Grantham to Creech-Jones, 3 May 1949.

78. CO 537/6032, savingram from Grantham to Griffiths, 31 March 1950. See also other correspondence from Hong Kong and minutes in the same file, which deal mainly with the incident.

79. CO 537/6032, savingram from Grantham to Griffiths, 31 March 1950, Grantham to Creech-Jones, 9 and 31 January 1950; CO 131/123, minutes of Executive Council, 31 January, and 7 and 14 February 1950; *Hong Kong Labour Report (1950–51)*, pp. 18–20 and 50–1; and CO 537/5303, CPIS (1950), No. 4.

80. *Hong Kong Labour Report (1949–50)*, pp. 50–1; *WKYP* leaders, 23 December 1949 and 10 February 1950; *KSJP* leader, 1 February 1950; *STJP* leader, 1 February 1950; *WHP* leader, 31 January 1950; and *TKP* leader, 1 February 1950.

81. *Hong Kong Labour Report (1950–51)*, pp. 30–4; and CO 537/6032, Grantham to Creech-Jones, 9 January 1950. For the 1925–6 strike, see P. Gillingham, *At the Peak* (Hong Kong, Macmillan and Co., 1983), pp. 33–46.

82. CO 537/6075, Hong Kong Police Special Branch (Monthly) Summaries, June–November 1950; CO 537/5306, CPIS (1950), No. 7; and *HSWP*, 18 August 1950, in *CPR 1950*.

83. CAB 128/17, CM 4(50)5, 7 February 1950; and CO 537/6032, savingram from Grantham to Griffiths, 31 March 1950.

84. See Chapter 4, pp. 86–7.

85. Consul-General (Hong Kong) to (U.S.) Secretary of State, 11 August and 3 November 1949, *FRUS*, Vol. VIII, 1949, pp. 480–1 and 578; *HSP*, 30 August 1949, in *CPR 1949*; FO 371/83260, FC10112/25G, report on Chinese Communist activities ending 31 December 1949.

86. FO 371/75782, F10305/1016/10G, savingram from Grantham to Creech-Jones, 27 June 1949 (including enclosure), and minute of Burgess, 13 July 1949.

87. Grantham, *Via Ports*, p. 163; CAB 128/17, CM 19(50)2, 6 April 1950, CM 24(50)6, 24 April 1950, and CM 40(50)7, 26 June 1950; CAB 129/39, CP (50)61, 3 April 1950, and CP (50)74, 21 April 1950. For details, see CO 537/4841 and CO 537/5628–33.

88. CO 537/5628, Younger (FO) to Hutchison (Peking), 6 March 1950; CO 537/5629, note of meeting with K. C. Potter and P. Chen (counsels of the Chinese government), 21 March 1950, and Griffiths to Grantham, 14 April 1950; CO 537/5630, opinions of the Law Officers, 3 April 1950; CO 537/5632, Grantham to Griffiths, 12 May 1950; and CO 537/5633, Hutchison to Bevin, 9, 11, 19, and 29 May, 28 July, and 30 October 1950.

89. Chou worked as a senior member of the editorial staff in the newspaper during this period, but was later returned to the Chinese mainland because he was considered to be a Western spy. He was jailed for four years and 're-educated' by the Communists before being sent to Hong Kong again to serve as an agent of the Chinese people. He left Hong Kong secretly and defected to Britain in 1961.

90. Eric Chou, *A Man Must Choose* (London, Longman, 1963), pp. 203–4.

91. FO 371/83263, FC 10112/80, savingram from Nicoll to Griffiths, 21 July 1950; CO 537/6075, Hong Kong Police Special Branch Report, July 1950; and FO 371/83525, FC 1671/10, letter from Seager (CO) to Trench (FO), 6 July 1950 (including enclosure).

92. FO 371/83263, FC 10112/80, dispatch from Nicoll to Griffiths, 21 July 1950; Cater interview (9 December 1982); Catron, *China and Hong Kong*, pp. 102 and 145–9; *Hong Kong Annual Report (1950)*, pp. 49–53; and CO 131/123, minute of Executive Council, 3 January 1950.

93. R. Deacon, *A History of the Chinese Secret Service* (London, Frederick Muller, 1974), pp. 428–31; and Chou, *A Man Must Choose*, p. 177.

94. The information was given by L. K. Tao (alias Tao Meng-huo, Director of Academia Sinica) who represented Lo in the meeting with the American Consul-General. Consul-General (Peping) to Secretary of State, 27 September 1949, *FRUS*, Vol. VIII, 1949, p. 539.

95. FO 371/75877, F10527/1192/10G, draft paper for J.I.C. Committee and minute of Dening, 25 July 1949; and FO 371/83263, dispatch from Nicoll to Griffiths, 21 July 1950.

96. FO 371/75783, F12541/1016/10, note of Colonial Office, 12 August 1949; CO 537/4835, extract from Grantham to Creech-Jones, 4 June 1949; and FO 371/75839, F12722/1061/10, Stevenson to Bevin, 25 August 1949.

97. FO 371/75781, F9880/1016/10, minute of Burgess, 7 July 1949; FO 371/83260, FC 10112/25G, minute of Burgess, 21 March 1950 and report on Chinese Communist activities for period ending 31 December 1949; and FO 371/83263, FC 10112/80, dispatch from Nicoll to Griffiths, 21 July 1950.

98. CO 537/6069, Grantham to Griffiths, 31 May 1950.

99. CO 537/6067, Grantham to Griffiths, 4 May and 6 June 1950.

100. CO 537/6542, enclosure to letter from Grantham to Paskin, 19 October 1949.

101. FO 371/75806, F3662/10129/10G, dispatch from Grantham to Creech-Jones, 24 February 1949, and F12048/10129/10G, dispatch from Grantham to Creech-Jones, 3 August 1949; and FO 371/83260, dispatch from Grantham to Creech-Jones, 1 February 1950.

102. CO 1023/101, savingram (with enclosure) from Grantham to Lyttelton (Secretary of State), 7 February 1952; CO 537/6075, Hong Kong Police Special Branch Summary, August 1950; and Chang Sheng, *Hsiang-chiang heh-shê-hui huo-tung chên-hsiang (The Real Story of the Activities of the Triad Societies in Hong Kong* (Hong Kong, Cosmos Books Limited, 1983), pp. 67–70 and 227–62. For the Kuomintang-Communist confrontation in October 1956, see *Report on the Riots in Kowloon and Tsuen Wan, October 10th to 12th, 1956* (Hong Kong, Government Printer, 1957).

103. K. P. Wu, *The Story of the Anti-Communistic Activities of the Patriotic Refugees now at Rennie's Mill Camp, Hong Kong* (in Chinese, but English subtitle is original) (Taipei, Universal Voice Press, 1958), pp. 103 and 107–23; FO 371/83263, FC 10112/73/G, dispatch from Nicoll to Griffiths, 17 July 1950; and Hambro, *Chinese Refugees*, pp. 64 and 176.

104. For the best analysis of the refugee problem, see Hambro, *Chinese Refugees*.

105. FO 371/83525, FC 1671/10, letter from Seager to Trench, 6 July 1950; FO 371/75806, F12048/10129/10G, dispatch from Grantham to Creech-Jones, 3 August 1949; and FO 371/83263, FC 10112/73/G, dispatch from Nicoll to Griffiths, 17 July 1950.

106. FO 371/75806, F12048/10129/10G, dispatch from Grantham to Creech-Jones, 3 August 1949; and FO 371/83260, FC 10112/15G, dispatch from Grantham to Creech-Jones, 1 February 1950.

107. FO 371/75806, F12048/10129/10G, dispatch from Grantham to Creech-Jones, 3 August 1949; FO 371/83260, FC 10112/15G, dispatch from Grantham to Creech-Jones, 1 February 1950; and CO 537/6542, Grantham to Creech-Jones, 20 January 1950.

108. For details of the proposals, see pp. 147–50.

109. *KSJP* editorials, 2 and 15 July 1949; *TKP* editorials, 6 and 20 July 1949; *WHP* editorial, 15 July 1949; and *WKYP* editorial, 15 July 1949.

110. For an overall analysis of Grantham's attitude, see Chapter 7, particularly pp. 187–90.

111. *Hong Kong Statistics (1947–1967)*, pp. 48, 79, and 88.

112. H. Ingrams, *Hong Kong* (London, HMSO, 1952), pp. 281–2; and Bernacchi interview (29 February 1984).

113. British traders did ask for representative government in the 1890s; their proposals were rejected by the British government on the grounds that they only asked to replace the benevolent autocracy of colonial officials by an oligarchy of British traders. See Endacott, *Government and People*, pp. 119–25. For the earliest idea of setting up a municipal council for Hong Kong, see Endacott, 'Proposals for Municipal Government in Early Hong Kong', *JOS*, Vol. 3, 1956, pp. 75–82.

114. HKGCC, *Report for 1950*, pp. 9 and 61; CAPMC 1949, circulars, 8 and 9

June 1949, and bulletin, 20 July 1949; CAPMC 1950, bulletins, 20 November and 20 December 1950; and *Hong Kong Statistics*, pp. 88 and 98.

115. *STJP*, 25 July and 28 September 1950, *CPR 1950*; and *FEER*, Vol. VIII, 1950, p. 766.

116. For a glimpse of the Chinese-language press coverage on the subject in English, see various translated editorials of December 1950 in *CPR 1950*, and January–February 1951 in *CPR 1951*.

117. According to the Chinese Reform Association's *Thirtieth Anniversary Special Publication (1949–1979)*, published in 1979, the Association was not formally established until 31 August 1949, although it had acted as a public body since May.

118. Hong Kong Reform Club, *Memorandum and Articles of Association of the Reform Club of Hong Kong* (Hong Kong, 1949), pp. 5–8; and Bernacchi interview.

119. Bernacchi interview; *SCMP* 23 June 1949; and CO 882/31, dispatch from Governor to Secretary of State, 25 August 1949.

120. Bernacchi interview.

121. Chen and Bernacchi interviews; *STJP*, 7 June 1949; and Hong Kong Chinese Reform Association, *Thirtieth Anniversary Special Publication (1949–1979)*, pp. 1–3. It is worthwhile noting that this publication makes no reference to any of the activities of the Association prior to 3 May 1953.

122. CO 537/6047, minute of Sidebotham, 4 October 1950, and note on the Hong Kong Chinese Reform Association, 9 November 1951.

123. The British government had of course never suggested that Hong Kong had to choose between reform of the legislature and the formation of a municipal council. The idea was that Hong Kong should make up its mind on what it wanted before the British government made a final decision.

124. *STJP*, 4, 5, and 6 May 1949; *KSJP*, 1, 5, and 7 July 1949; and CO 882/31, dispatch from Governor to Secretary of State, 25 August 1949.

125. *KSJP* 17, 19, 22, 23, and 30 June 1949, and 1, 2, 5, 6, and 8 July 1949.

126. *KSJP* 8, 11, 14, and 15 July 1949; *WKYP* editorial, 15 July 1949; *TKP* editorial, 20 July 1949; *WHP* editorial, 15 July 1949; and CO 537/6047, letter from Chinese Reform Association to Sir H. Poynton, 20 July 1950.

127. See CO 537/6046, minutes of Sidebotham, 11 January 1950 and 13 April 1950.

128. *SCMP*, 2 and 3 November 1949 (editorial); *WKYP* editorial, 22 February 1950; and *FEER*, Vol. VIII, 1950, p. 264 and issue dated 27 April 1950.

Notes to Chapter 6

1. CO 1023/41, note of a meeting (with Nicoll) in Paskin's room, 14 February 1952, and minute of Paskin, 10 March 1952.

2. CO 131/146, minute of Executive Council, 20 November 1951; and CO 1023/41, note of a discussion with the Governor of Hong Kong, 10 July 1952. Carrie was appointed Special Adviser in November 1948 to take over the duties of his predecessors, Lee and Hazlerigg. See Chapter 4 for details.

3. CO 1023/41, minute of Sidebotham, 25 January 1952. The quotation is cited by Sidebotham from the original record of the meeting.

4. Lord Chandos, *The Memoirs of Lord Chandos* (London, The Bodley Head, 1962), p. 376; and CO 1023/41, minute of Hall, 24 January 1952. The quotation was taken by Hall from the original record of the meeting.

5. *SCMP*, 15 December 1951, and *KSJP*, 12 and 15 December 1951.

6. *SCMP*, 13 December 1951, and *WKYP*, 13 December 1951.

7. *SCMP* editorials, 13 and 14 December 1951, and *WKYP*, *KSJP*, and *STJP*, 12-16 December 1951.

8. *WKYP* editorial, 12 December 1951, *STJP* editorial, 13 December 1951, and *SCMP*, 13 December 1951.

9. *SCMP* letters to the editor, 12, 13, and 14 December 1951.

10. CO 1023/41, minute of Paskin, 10 March 1952, and letter from Grantham to Paskin, 8 January 1952; and CO 131/146, minute of Executive Council, 18 December 1951.

11. CO 1023/41, minutes of Hall, 24 January 1952, and Sidebotham, 25 January and 15 February 1952.

12. CO 1023/41, minute of Sidebotham, 25 January 1952; and *FEER*, Vol. XII, 1952, pp. 1 and 100. The relevant political intelligence for 1951 is in six files: CO 537/6797-802. See also Chapter 5 for the differences in views between the Reform Club and the HKCGCC in 1949.

13. CO 1023/41, note of a meeting (with Nicoll) in Paskin's room, 14 February 1952. Telegraphic summaries of Chinese propaganda for this period are in CO 537/7120.

14. CO 1023/41, minutes of Sidebotham, 16 February 1952, Paskin, 10 March 1952, and Mackintosh (Lyttelton's Private Secretary), 19 March 1952.

15. FO 371/99251, FC 10118/1, letter from Sidebotham to Johnson, 19 March 1952; minute of Oakeshott, 31 March 1952; and letter from Scott to Sidebotham, 31 March 1952; CO 1023/41, minute of Anderson.

16. CAB 128/25, CC 54(52)6, 20 May 1952; and CAB 129/52, C (52)165, 16 May 1952.

17. CO 131/147, minutes of Executive Council, 25 March and 1 April 1952; and *Hong Kong Hansard 1952*, minute of Legislative Council, 9 April 1952.

18. *SCMP* news, 30 May 1952.

19. *SCMP* news and editorial, 30 May 1952; and Chen interview (4 May 1984). For Chen's connection with the Chinese Communists, see P. Chen, *China Called Me: My Life inside the Chinese Revolution* (Boston, Little, Brown & Co., 1979).

20. *STJP* editorial, 24 April 1952; and *SCMP* letters to the editor, 24 May 1952 (from Lam), and 27 May 1952 (from W.), and news, 31 May 1952.

21. *SCMP* news and editorials, 31 May and 1 June 1952.

22. *STJP* editorial, 1 June 1952.

23. *SCMP* editorials, 29 and 30 May 1952; and CO 1023/41, Grantham to Lyttelton, 14 June 1952.

24. *SCMP* news and editorials, 31 May and 1 June 1952.

25. CO 1023/41, Grantham to Lyttelton, 14 June 1952.

26. CO 131/147, minutes of Executive Council, 17 and 24 June 1952; and CO 1023/41, Grantham to Lyttelton (for Sidebotham), 26 June 1952, and note of a discussion with the Governor of Hong Kong, 10 July 1952.

27. CO 131/147, minutes of Executive Council, 17 and 24 June 1952; and CO 1023/41, Grantham to Lyttelton (for Sidebotham), 26 June 1952, and note of a discussion with the Governor of Hong Kong, 10 July 1952.

28. See pp. 115-17.

29. CO 1023/41, minutes of Paskin, 11 July 1952, and Jeffries, 12 July 1952; and Grantham, *Via Ports*, p. 112.

30. CO 1023/41, Officer Administering the Government to Secretary of State (for Grantham), 10 September 1952, and extract from monthly report by the Trade Commissioner in Hong Kong; and *STJP*, 19 June 1952.

31. CAB 128/25, CC 80(52)3, 18 September 1952; and CAB 129/54, C (52)298, 11 September 1952.

32. CO 131/148, minutes of Executive Council, 9 September and 14 and 21 October 1952; CO 1023/41, Officer Administering the Government to Secretary

of State (for Grantham), 10 September 1952, and minute of Sidebotham, 13 October 1952; 505 *HC DEB* 5s, c. 70; and *Hong Kong Hansard 1952*, p. 252.

33. *SCMP* editorial, 22 October 1952; *FEER*, Vol. XIII, 1952, p. 556; *STJP* editorial, 23 October 1952; and CO 1023/41, savingram from Grantham to Lyttelton, 12 November 1952.

34. CO 131/147, minute of Executive Council, 4 March 1952; CO 131/148, minute of Executive Council, 16 September 1952; and *STJP*, 21 September 1952, in *CPR 1952*.

35. *Hong Kong Annual Report 1952*, p.11.

36. CO 1023/41, Officer Administering the Government to Secretary of State (for Grantham), 10 September 1952; letters between Bernacchi and Black (Colonial Secretary, Hong Kong), October and November 1952; savingram from Grantham to Lyttelton, 12 November 1952; dispatches from Grantham to Lyttelton, 28 October and 6 November 1953; and Reuters report, 18 December 1953; CO 131/149, minutes of Executive Council, 24 February and 3 and 24 March 1953; and CO 131/150, minutes of Executive Council, 1 September and 13 October 1953.

37. *Hong Kong Annual Report 1951*, pp. 8 and 43-7; *Hong Kong Annual Report 1952*, tables between pp. 54-5; HKGCC, *Report for 1951*, pp. 9 and 12-14; *FEER*, Vol. X, 1951, pp. 169, 299, and 309-10; *FEER*, Vol. XI, 1951, pp. 1 and 840-1; *FEER*, Vol. XII, 1952, pp. 332-4, 345, 541-3, and 601-3; and *Wah Kiu Year Book 1952*, Section 6, p. 13.

38. *Hong Kong Annual Report 1952*, pp. 11-14 and 55-8; HKGCC, *Report for 1953*, pp. 59-63; *Hong Kong Statistics*, pp. 88 and 98; FO 371/105188, FC 1011, dispatch from Lamb (Peking) to Eden, 18 February 1953; *FEER*, Vol. XII, 1952, pp. 601-3; and *Wah Kiu Year Book 1953*, Section 6, pp. 21-2.

39. CO 537/7643, letter from Clarke to Hall, 18 April 1952; HKGCC, *Report for 1952*, pp. 9-10 and 48-9; HKGCC, *Report for 1953*, pp. 59-61; and *Hong Kong Statistics*, p. 88.

40. CAPMC 1951, British Chamber (Shanghai) to China Association, confidential telegram No. 44 (1951); CAPMC 1952, Consul-General (Shanghai) to Peking, 11 June 1952, and Bulletin No. 74 (20 July 1952); FO 371/105188, FC1011, dispatch from Lamb to Eden, 18 February 1953; CO 537/7643, letter from Clarke to Hall, 18 April 1952; and R. Boardman, *Britain and the People's Republic of China: 1949-1974* (London and Basingstoke, Macmillan and Co., 1976), pp. 81-3. For a detailed description of how British traders were forced to leave Shanghai after the Communist take-over in 1949, see N. Barber, *The Fall of Shanghai* (New York, Coward, McCann & Geoghegan, 1979), pp. 160-236.

41. *Wah Kiu Year Book 1953*, Section 6, pp. 21-2.

42. Most Chinese residents of Hong Kong at that time, particularly those who did not know the English language, did not distinguish between the Americans and the British. Ironically, it was the local Communists who were (relatively) most aware of the great gap separating the Americans from the British in Far Eastern affairs.

43. CO 537/6797, *CPIS*, January 1951; *STJP* editorial, 11 January 1951; *WKYP* editorial, 11 January 1951; *TKP* editorial, 10 January 1951; and *WHP*, 12 January 1951, in *CPR 1951*; and *Wah Kiu Year Book 1952*, Section 5, p. 3.

44. CO 537/7333, letter from Scott to Paskin, 17 July 1951, and savingram from Grantham to Griffiths, 19 September 1951.

45. FO 371/99243, FC 10111/3, memorandum from Grantham for the Prime Minister (undated, but submitted in January 1952); Grantham, *Via Ports*, pp. 153-9; and *Police Report 1951-2*, p. 49.

46. CO 537/6798, *CPIS*, March 1951.

47. CO 537/7120, enclosure to savingram from Grantham to Lyttelton, 22

October 1951 (Panikkar interview). British diplomats' periodic reports on the themes of Communist propaganda for 1951 are in three files: CO 537/7118-20. For the Foreign Office's view of Anglo-Chinese relations in this period, see FO 371/75827, F18896/1023/10, 'Sino-British Exchanges on the Question of Establishing Full Diplomatic Representation', 18 June 1970 [sic].

48. FO 371/99260, FC 1015/4, Lamb to Eden, 1 February 1952; and FC1025/7/G, undated brief of Scott for Eden; FO 371/99243, FC10111/17, Lamb to Eden, 26 January 1952; and WHP editorial, 31 January 1952, in CPR 1952.

49. CAB 128/19, CM 23(51)4, 2 April 1952, and CM 24(51)2, 5 April 1952; Grantham, Via Ports, pp. 163-4; and CO 537/7119, Lamb to Morrison, 4 and 30 May 1951.

50. WHP editorials, 21 and 31 January, 7, 15, and 17 February, and 10 and 12 March 1952; WHP news, 5 March 1952; TKP news, 6 February and 6 March 1952; and TKP editorial, 11 February and 10 and 12 March 1952, in CPR 1952; FO 371/99243, FC10111/7, Grantham to Lyttelton, 10 and 16 January 1952; FC10111/8, Lamb to Eden, 17 January 1952; and FC10111/14, Lamb to Eden, 26 January 1952; FO 371/99244, FC10111/30, minute of Oakeshot, 9 April 1952; FC10111/39, Lamb to Eden, 8 March 1952; and FC10111/58, Lamb to Eden, 11 May 1952; and CO 131/147, minutes of Executive Council, 8, 15, and 22 January, 11 and 18 March, 1 and 8 April, and 6 May 1952.

51. Lien ho pao (Canton) editorial, 5 February 1952, quoted in TKP, 6 February 1952, in CPR 1952.

52. FO371/99244, FC10111/36, savingram from Grantham to Lyttelton, 7 February 1952; and FC10111/50, minute of Oakeshot, 9 April 1952; and FO371/99245, FC10111/74, letter from Aldington (Political Adviser, Hong Kong) to Johnston (F.O.), 3 June 1952. For the newspaper trial, see Ta Kung Pao an t'ing shen chi (An Account of the Ta Kung Pao Trial) (Hong Kong, Lien Yu Chu-pan-shê, undated, probably 1952).

53. Police Report 1951-52, p. 47; CO 131/147, minutes of Executive Council, 15 January and 19 and 28 February 1952; FO371/99243, FC10111/33, all telegrams and minutes; and TKP news, 2 and 6 March, and WHP news, 1 and 5 March 1952, in CPR 1952.

54. CO 131/148, minutes of Executive Council, 15, 22, and 29 July, 5 and 12 August, and 9 and 16 September 1952; and TKP and WHP news and editorials, 28 July-4 August and 27-30 September 1952.

55. CO 1023/101, Grantham to Lyttelton, 14 October 1952; minute of Littler, 14 October 1952; letter from Aldington to Jacobs-Larkcom (British Consul, Tamsui, Taiwan), 27 November 1952; and dispatch from Grantham to Lyttelton, 28 April 1953; and KSJP, 2 October 1952, TKP, 2-3 October 1952, WHP, 2-3 October 1952, WKYP, 4 October 1952, and STJP, 11 October 1952, in CPR 1952.

56. CO 1023/101, letter from Aldington to Johnston, 4 November 1952.

57. CO 1023/101, letters from Aldington to Johnston, 4 November 1952, and from Aldington to Briggs (British Consul, Tamsui, Taiwan), 5 August 1953.

58. FEER, Vol. XIII, 1952, pp. 114-15. For the Kuomintang's guerrilla activities, see Chapter 5, pp. 136-7.

59. Police Report 1951-52, p. 49.

Notes to Chapter 7

1. CO 537/1651, confidential dispatch from Young to Creech-Jones, 22 October 1946; and CO 537/2188, minute of Galsworthy, 24 June 1947. See also Chapter 3.

2. It must be emphasized again that at that stage Young was merely thinking of leading Hong Kong to attain representative or reponsible government within the Empire, rather than independence.

3. Grantham, *Via Ports*, pp. 111 and 172; and Grantham, 'Hong Kong', *JRCAS*, Vol. 46, p. 121.

4. For the importance of the changes in the socio-economic and political situation as a factor which affected the decisions of the two governors, see pp. 201–8.

5. Todd interview.

6. CO 537/5400, minute of ninth STC meeting, 12 July 1950 (item 4).

7. Wight, *The Development of the Legislative Council*, pp. 74 and 77–8; and Todd interview.

8. Grantham, *Via Ports*, pp. 3–19, particularly p. 14.

9. Enclosure to a letter to the author from Bailey (Senior Crown Counsel, Hong Kong), 1 September 1986.

10. Mss 940.547252 G49.

11. Mss 940.547252 G49; Gimson's diary, 5 January 1945, Mss Ind. Ocn. s. 222; and Mss 940.54252 G49i. See also Chapter 2.

12. For the attitude, training, and background of the Hong Kong Cadets before 1945, see Lethbridge, *Stability and Change*, particularly p. 39.

13. CO 325/42/55104/2, memorandum of Smith, 2 August 1942.

14. MacDougall interview; CO 825/35/55104, minutes and memoranda of MacDougall and Gent (both individually and jointly), June and July 1942; FO371/35905, letters from MacDougall to Sabine, 22 and 30 December 1942; and Lee and Petter, *The Colonial Office*, pp. 135–6 and 142.

15. This was the usual practice. See C. J. Jeffries, *The Colonial Office* (London, Allen and Unwin, 1956), pp. 35–7.

16. As shown in Chapter 6, Sidebotham alone among the senior officials seriously doubted the wisdom of Grantham's recommendation to proceed with the reforms in early 1952 because information available to him was incomplete.

17. Paskin interview.

18. Cmd. 7709, p. 51; Mss Brit. Emp. s. 332 (Creech-Jones Papers), Box 4, File 4, ff. 9–10, and Box 9, File 8, Item 18. See also Pearce, *The Turning Point*, pp. 100–4.

19. Goldsworthy, *Colonial Issues*, p. 22.

20. Goldsworthy, *Colonial Issues*, p. 15.

21. Todd interview.

22. Todd interview.

23. Grantham, *Via Ports*, p. 110.

24. Tang, C. C., *Social Leaders in Hong Kong and Macau* (Hong Kong, Hong Kong Associated Press, 1958), pp. 2–6; and *The Hong Kong Directory 1949* (Hong Kong, The Newspaper Enterprise Ltd., 1950), pp. 643, 652, 692, 755, 784–5, and 888.

25. C. B. Burgess, 'Great Britain and Constitutional Developments in Hong Kong, 1945–1952' (paper deposited at Rhodes House Library), p. 8.

26. Burgess, 'Great Britain and Constitutional Developments in Hong Kong', p. 7.

27. *Hong Kong Hansard 1949*, p. 191.

28. Grantham's 1950 proposals are not relevant here because they did not provide for direct elections and were in any case not made public. For political participation in Hong Kong, see also J. S. Hoadley, 'Political Participation of Hong Kong Chinese: Patterns and Trends', *Asian Survey*, Vol. 13, 1973, pp. 604–16. Readers are reminded that Hoadley's research was carried out towards the end of the 1960s, and his findings are no longer up to date.

29. See, for example, CO 882/31, secret dispatch (Part II) from Grantham to Creech-Jones, 25 August 1949.

30. CO 882/31, secret dispatch (Part II) from Grantham to Creech-Jones, 25 August 1949.

31. CO 537/1651, confidential dispatch from Young to Creech-Jones, 22 October 1946.

32. CO 131/116, minute of Executive Council, 2 August 1946.

33. CO 537/2189, extract from letter from Hazlerigg to Roberts-Wray, 24 November 1947. For the danger of Communist infiltration, see pp. 205–6.

34. Deacon even classified Ko (romanized as C. H. Kao by Deacon) as a Communist 'fat cat' in Hong Kong. Deacon, *Chinese Secret Service*, p. 432.

35. See CO 537/2193, savingram from Grantham to Creech-Jones, October–December 1947; *HSWP* editorial, 3 October 1947; and *STJP* editorial, 17 October 1947. For a discussion of Communist infiltration, see pp. 205–6.

36. FO 371/83260, FC10112/15G, top-secret dispatch from Grantham to Creech-Jones, 1 February 1950.

37. CO 537/3703, minutes of Wallace, 9 November 1948, and Sidebotham, 10 November 1948.

38. Chen interview; and *SCMP*, 30 May 1952.

39. Chou, *A Man Must Choose*, p. 210.

40. CO 882/31, secret dispatch (Part III) from Grantham to Creech-Jones, 25 August 1949.

41. Nevertheless, it is possible that during the excesses of the 1967 riots — which were ripples of the Cultural Revolution inside China — the local Communists might have attempted to harass directly elected municipal councillors, had there been any. This does not necessarily mean, however, that the councillors would have kowtowed to the Communists.

Notes to Chapter 8

1. Grantham, *Via Ports*, p. 105.

2. CO 967/42, minute of the Cabinet Commonwealth Affairs Committee, 29 October 1948.

3. For a good and concise exposition of the more usual form of constitutional evolution in British colonies, see A. F. Madden, 'How Colonies Grow into Dominions', *United Empire*, Vol. 41, No. 3, 1950, pp. 159–64.

4. *Report of the Working Party on Local Administration* (Hong Kong, Government Printer, 1966), p. 2.

5. *Hong Kong 1982* (Hong Kong, Government Printer, 1982), p. 241. It is worthy of note that this statement is deleted in the subsequent issue of the year-book. See *Hong Kong 1983* (Hong Kong, Government Printer, 1983), particularly pp. 252–68.

Bibliography

Unpublished Primary Sources

1. Official documents (Public Records Office, Kew)
   Cabinet conclusions, CAB 128.
   Cabinet memoranda, CAB 129.
   Cabinet, Far Eastern (Official) Committee: minutes and papers, CAB 134.
   Colonial Office, Original Correspondence — Hong Kong (up to and including 1951), CO 129.
   Colonial Office, Hong Kong Executive Council minutes, CO 131.
   Colonial Office, Original Correspondence — supplementary (up to and including 1951), CO 537.
   Colonial Office, Original Correspondence — Eastern, CO 825.
   Colonial Office, Confidential Prints — Eastern, No. 181 (Constitution of Hong Kong: Correspondence 1946–1952), CO 882.
   Colonial Office, Private Office Papers, CO 967.
   Colonial Office, Original Correspondence — Hong Kong and Pacific (since 1952), CO 1023.
   Chiefs of Staff Committee, memoranda, DEFE 5.
   Chiefs of Staff Committee, Reports of the Joint Planning Staff, DEFE 6.
   Foreign Office, Far Eastern Department General Political Files, FO 371.

2. Official documents (Public Records Office of Hong Kong)
   Colonial Secretariat, General Correspondence (General Branch), HKRS 41.
   Colonial Secretariat, General Correspondence (British Military Administration), HKRS 169.
   Colonial Secretariat, General Correspondence (Civil Affairs Administration), HKRS 170.
   Colonial Secretariat, General Correspondence (Hong Kong Planning Unit), HKRS 211.

3. Private papers and manuscripts
   (a) Rhodes House Library, Oxford:
       Burgess, Claude (Great Britain and Constitutional Developments in Hong Kong 1945–1952).
       Creech-Jones, Arthur (reference: Mss Brit. Emp. s. 332, Boxes 4, 9, 16, 26, and 57).

Dowbiggin, Colonel H. B. L. (Hong Kong Reminiscences 1906–1965) (reference: Mss Ind. Ocn. s. 151).

Fabian Colonial Bureau (reference: Mss Brit. Emp. s. 365, Box 152).

Gimson, Sir Franklin (reference: Mss Ind. Ocn. s. 222).

Grantham, Sir Alexander (reference: Mss Brit. Emp. s. 288).

Forster, L. (reference: Mss Ind. Ocn. s. 177).

(b) Hung On-to Memorial Library, University of Hong Kong:

Gimson, Sir Franklin (reference: Mss 940.547252 G49 (Internment of European Civilians at Stanley during the Japanese Occupation of Hong Kong, 1941–1945); and reference: Mss 940.547252 G49i (Hong Kong Reclaimed)).

(c) Others:

China Association, London (School of Oriental and African Studies, London).

Herklots, Dr G. A. C. (in Dr Herklots' possession).

Swire, John (Diary of Post-war Trips to China and Australia) (in possession of Sir Adrian Swire).

Published Primary Sources

1. Command Papers

Cmd. 7167: The Colonial Empire (1939–1947) (London, HMSO, 1947).

Cmd. 7433: The Colonial Empire (1947–1948) (London, HMSO, 1948).

Cmd. 7709: British Dependencies in the Far East (1945–1949) (London, HMSO, 1949).

Cmd. 7715: The Colonial Territories (1948–1949) (London, HMSO, 1949).

Cmd. 7958: The Colonial Territories (1949–1950) (London, HMSO, 1950).

Cmd. 8243: The Colonial Territories (1950–1951) (London, HMSO, 1951).

Cmd. 8553: The Colonial Territories (1951–1952) (London, HMSO, 1952).

2. Other British government publications

The Colonial Office List (1945–1952).

Hansard, Parliamentary Debates, 1945–1952.

3. Hong Kong government publications

Annual Departmental Reports by the Commissioner of Labour (1946–1952) (Hong Kong, Government Printer, 1946–52).

Annual Departmental Reports by the Commissioner of Police (1946–1952) (Hong Kong, Government Printer, 1946–52).

British Military Administration Gazette (1945–1946) (Hong Kong, Government Printer, 1945–6).

The British Military Administration of Hong Kong, August 1945 to April 1946 (Hong Kong, Government Printer, 1946).

Civil Service List for 1950–1951 (Hong Kong, Government Printer, 1951).

Government Departments in Hong Kong, 1841–1969 (compiled by G. C. Hamilton) (Hong Kong, Government Printer, 1969).

Hong Kong (compiled by Hong Kong Public Relations Office) (Hong Kong, Government Printer, 1952).

Hong Kong 1982 (Hong Kong, Government Printer, 1982).

Hong Kong 1983 (Hong Kong, Government Printer, 1983).

Hong Kong 1984 (Hong Kong, Government Printer, 1984).

Hong Kong 1985 (Hong Kong, Government Printer, 1985).

Hong Kong 1986 (Hong Kong, Government Printer, 1986).

Hong Kong Annual Reports (1946–1952) (Hong Kong, Government Printer, 1946–52).

Hong Kong Government Gazette (1946–1952) (Hong Kong, Government Printer, 1946–52).

Hong Kong Hansard, Reports of the Sittings of the Legislative Council of Hong Kong, 1946–1952 (Hong Kong, Government Printer, 1946–52).

Hong Kong Statistics, 1947–1967 (compiled by Census and Statistics Department) (Hong Kong, Government Printer, 1969).

Report on the Riots in Kowloon and Tsuen Wan October 10th to 12th, 1956, together with Covering Despatch dated 23rd December 1956, from the Governor of Hong Kong to the Secretary of State for the Colonies (Hong Kong, Government Printer, 1957).

Report of the Working Party on Local Administration (Hong Kong, Government Printer, 1966).

Staff Biographies, Hong Kong Government (1974).

The Urban Council, 1883–1983 (Hong Kong, 1983).

Urban Council Report on the Reform of Local Government (Hong Kong, Government Printer, 1969).

4. United States government publications

Foreign Relations of the United States 1942: China (Washington, US Government Printer, 1956).

Foreign Relations of the United States 1943: China (Washington, US Government Printer, 1957).

Foreign Relations of the United States 1944: China (Washington, US Government Printer 1967).

Foreign Relations of the United States 1945: China (Washington, US Government Printer 1969).

Foreign Relations of the United States 1947: China (Washington, US Government Printer 1972).

Foreign Relations of the United States 1948: China (Vols. 7 and 8, Washington, US Government Printer, 1973).

Foreign Relations of the United States 1949: China (Vols. 8 and 9, Washington, US Government Printer, 1974).

Foreign Relations of the United States 1950: China (Washington, US Government Printer, 1976).

Foreign Relations of the United States 1951: Asia and Pacific (Washington, US Government Printer, 1977).

5. English-language newspapers and periodicals
   Far Eastern Economic Review (Hong Kong).
   Manchester Guardian (newspaper cuttings of the Fabian Colonial Bureau), Rhodes House Library, Mss Brit. Emp. s. 365, Box 152.
   South China Morning Post (Hong Kong).
   The Times (newspaper cuttings of the Fabian Colonial Bureau), Rhodes House Library, Mss Brit. Emp. s. 365, Box 152.

6. Hong Kong Chinese-language newspapers
   Hong Kong shih pao (香港時報).
   Hsin sheng wan pao (新生晚報).
   Hwa shiang pao (華商報).
   Kung shang jih pao (工商日報).
   Sing tao jih pao (星島日報).
   Ta kung pao (大公報) (Hong Kong edition, since 1948).
   Wah kiu yat pao (華僑日報).
   Wen wei pao (文滙報) (Hong Kong edition, since 1949).

   A selected translation of Hong Kong Chinese-language newspapers since July 1947 is available in *Chinese Press Review* (prepared and privately published by the American Consulate-General, Hong Kong). Printed copies are available at the University Services Centre, Hong Kong; a micro-film version can be found at the Hung On-to Memorial Library, University of Hong Kong.

7. Pro-Communist Hong Kong Chinese-language publications
   Ching chi nien-pao (經濟年報) (1 January 1951).
   Chun chung (群衆), Nos. 131–43 (August–September 1949).
   Economic Newsletter (經濟導報), Vol. 3, No. 11 (27 March 1948); Vol. 3, No. 15 (4 May 1948).
   Hong Kong Chinese Reform Association (香港華人革新會), *Special Thirtieth Anniversary Publication of the Hong Kong Chinese*

Reform Association, 1949–1979 (香港華人革新會成立三十
周年特刊 1949–1979) (some entries in English) (1979).
Hwa Shiang Pao, *I-chiu-shih-chi sou-cha* (一九四七手冊) (1947).
——, *I-chiu-shih-pa sou-cha* (一九四八手冊) (1948).
Kuang Min Pao (光明報), Vol. 3, No. 10 (16 July 1949).
Life in Hong Kong (生活在香港) (original English subtitle), Vol. 1,
No. 1 (10 October 1947).
New Society Fortnightly (新社會半月刊) (original English subtitle),
No. 1 (1 November 1949).
Tai-ku chih-kung hui-kan: san-chou-nien chi-nien te-kan (太古職工
會刊：三周年紀念特刊) (10 April 1949).
Ta Kung Pao aw ting-shen-chi (大公報案聽審記) (undated, probably
1952).

8. Pro-Kuomintang Hong Kong Chinese-language publications
Chiang-chiu kung-tuan lien-ho-tsung-hui (港九工團聯合總會), *Hong
Kong Labour* (香港勞工) (original English subtitle), 1 January
1949 issue.
Chinese Engineers' Institute Bi-monthly (華機半月刊) (original English
subtitle), Vol. 1, No. 2, 1946.
Chung-kuo kuo-min-tang chu-chiang-ao tsung-chih-pu (中國國民黨駐
港澳總支部), *Shuang-shih te-kan* (Hong Kong, 1946). 雙十特刊
——, *Tang-yuan tung-hsin* (黨員通訊), No. 7 (21 June 1947); No. 8 (1
July 1947); joint issue Nos. 9–10 (21 July 1947); No. 12 (21
August 1947).
——, *Yuan-tan Te-kan* (元旦特刊) (Hong Kong, 1948).
Chung Pao (忠報) (weekly), Vol. 1, Nos. 1 and 2 (1 and 8 July 1947).
Radar (雷達) (original English subtitle), No. 1 (5 June 1946).
Sin Wen Tien Ti (新聞天地), No. 83 (17 September 1949); No. 95 (10
September 1949).
South Seas Weekly (南海周報) (original English subtitle), No. 26 (23
July 1947).

Oral Evidence

1. Officials
Burgess, Claude (15 June 1983).
Caine, Sir Sydney (15 April 1983).
Cater, Sir Jack (9 December 1982).
Cowperthwaite, Sir John (4 April 1983).
Herklots, Dr G. A. C. (18–19 January 1983).
MacDougall, D. M. (17 February 1983).
Paskin, Lady (née A. M. Ruston) (28 July 1983).
Sidebotham, J. B. (24 June 1983).
Todd, Reverend Alastair (11 July 1983).

2. Non-officials

Bernacchi, Brook (29 February 1984).

Chen, Percy (4 May 1984).

Leung, To (29 February 1984).

Ng, Wah (29 February 1984).

Secondary Sources

1. Books in English

Alderson, G. L. D., *History of Royal Air Force Kai Tak* (Hong Kong, Royal Air Force Kai Tak, 1972).

Allen, Louis, *The End of the War in Asia* (London, Hart-Davis, MacGibbon, 1976).

Association for Radical East Asian Studies, *Hong Kong: Britain's Last Colonial Stronghold* (London, the Association, 1972).

Attlee, Earl, *Empire into Commonwealth* (London, Oxford University Press, 1961).

Barber, Noel, *The Fall of Shanghai* (New York, Coward, McCann & Geoghegan, 1979).

Bernacchi, B. A., *Hong Kong Reform Club: Second Ten Years Anniversary Report (1958–1968)* (Hong Kong, Hong Kong Reform Club, 1969).

Birch, A. and Cole, M., *Captive Years* (Hong Kong, Heinemann Asia, 1982).

Blood, Sir Hilary, *The Future of the Smaller Commonwealth Territories* (London, British Commercial Union, 1967).

Boardman, Robert, *Britain and the People's Republic of China: 1949–1974* (London and Basingstoke, Macmillan & Co., 1976).

Bonavia, David, *Hong Kong 1997* (Hong Kong, Columbus Books, 1984).

Bowle, John, *The Imperial Achievement* (London, Secker and Warburg, 1974).

Braga, J. M. (compiled), *Hong Kong Business Symposium* (Hong Kong, South China Morning Post, 1957).

Bullock, Alan, *Ernest Bevin: Foreign Secretary* (New York and London, W. W. Norton & Co., 1983).

Burgess, C. B., *A Problem of People* (Hong Kong, Government Printer, 1966).

——, *Hong Kong's Image* (Hong Kong, Government Printer, 1962).

Carrington, C. E., *The Liquidation of the British Empire* (London, George C. Harrap & Co. Ltd., 1961).

Catron, G. W., *China and Hong Kong, 1945–1967* (unpublished Harvard University Ph.D. thesis, 1971) (copy at Hung On-to Memorial Library, Hong Kong).

Chandos, Lord, *The Memoirs of Lord Chandos* (London, The Bodley Head, 1962).

Chau, Lam-yan and Lau Siu-kai, *Development, Colonial Rule and Intergroup Politics in a Chinese Village in Hong Kong* (Hong Kong Chinese University occasional paper No. 90) (Hong Kong, The Chinese University of Hong Kong, 1980).

Chen, Percy, *China Called Me: My Life inside the Chinese Revolution* (Boston, Little, Brown & Co., 1979).

Cheong-leen, Hilton, *Hong Kong Tomorrow: A Collection of Speeches and Articles by Hilton Cheong-leen* (Hong Kong, 1962).

Chou, Eric, *A Man Must Choose* (London, Longman, 1963).

Chung Chi College, *1951–1976: A Quarter Century of Hong Kong* (Hong Kong, The Chinese University of Hong Kong, 1977).

Coates, Austin, *Myself a Mandarin: Memoirs of a Special Magistrate* (Hong Kong, Heinemann Asia, 1975).

Collins, Maurice, *Wayfoong: The Hong Kong and Shanghai Banking Corporation* (London, Faber & Faber Ltd., 1965).

Colonial Office and Central Office of Information, *Introducing the Eastern Dependencies* (London, HMSO, 1949).

Crisswell, C. N., *The Taipans: Hong Kong's Merchant Princes* (Hong Kong, Oxford University Press, 1981).

Dawson, R. M. (ed.), *The Development of Dominion Status 1900–1936* (Oxford, Clarendon Press, 1937).

Deacon, R., *A History of the Chinese Secret Service* (London, Frederick Muller, 1974).

Dial, O. E., *An Evaluation of the Impact of China's Refugees in Hong Kong on the Structure of the Colony's Government in the Period Following World War II* (unpublished Clarement Graduate School and University Centre Ph.D. thesis, 1965) (copy at Hung On-to Memorial Library, Hong Kong).

Donnison, F. S. V., *British Military Administration in the Far East: 1943–1946* (London, HMSO, 1956).

Earl, Lawrence, *Yangtze Incident* (London, London & Co., 1950).

Eastman, L. E., *China under Nationalist Rule 1927–1937* (Cambridge, Massachusetts, Harvard University Press, 1974).

Easey, Walter, *Ducking Responsibility* (Hong Kong Research Project Pamphlet No. 4) (Manchester, Christian Statesman, 1980).

Emerson, G. C., *Stanley Internment Camp, Hong Kong 1942–1945* (Hong Kong University M.Phil thesis, 1973).

Endacott, G. B., *Government and People in Hong Kong 1841–1962: A Constitutional History* (Hong Kong, Hong Kong University Press, 1964).

——, *A History of Hong Kong* (Hong Kong, Oxford University Press, 1964, second edition).

—— (ed.), *An Eastern Entrepôt: A Collection of Documents Illustrating the History of Hong Kong* (London, HMSO, 1964).

—— and Birch, A., *Hong Kong Eclipse* (Hong Kong, Oxford University Press, 1978).

Ewart, Kenneth, *Some Aspects of Local Government* (Bathurst, Gambia, V. E. Davies, 1946).

Fisher, S. F., *Eurasian in Hong Kong: A Sociological Study of a Marginal Group* (unpublished Hong Kong University M.Phil. thesis, 1975).

Fitzgerald, Stephen, *China and the Overseas Chinese* (Cambridge, Cambridge University Press, 1972).

Geddes, Philip, *In the Mouth of the Dragon* (London, Century, 1982).

Gillingham, Paul, *At the Peak* (Hong Kong, Macmillan & Co., 1983).

Gleason, Gene, *Hong Kong* (Hong Kong, Robert Hale Ltd., 1964).

Goldsworthy, David, *Colonial Issues in British Politics: 1945–1961* (Oxford, Oxford University Press, 1971).

Grantham, A. W. G. H., *Via Ports, From Hong Kong to Hong Kong* (Hong Kong, Hong Kong University Press, 1965).

Hailey, Lord, *Great Britain, India and the Colonial Developments in the Post-war World* (Toronto, University of Toronto Press, 1943).

Hambro, Edvard, *The Problems of Chinese Refugees in Hong Kong* (Leyden, A. J. Sijthoff, 1955).

Hamilton, G. C., *Government Departments in Hong Kong: 1841–1969* (Hong Kong, Government Printer, 1969).

Haqqi, S. A. H., *The Colonial Policy of the Labour Government (1945–1951)* (Aligarh, Muslim University, 1960).

Harrison, Brian (ed.), *University of Hong Kong: The First Fifty Years* (Hong Kong, Hong Kong University Press, 1962).

Hinden, Rita (ed.), *Fabian Colonial Essays* (London, George Allen & Unwin, 1945).

Ho, S. D-F., *A Hundred Years of Hong Kong* (unpublished Princeton University Ph.D. thesis, 1946) (copy at the Hung On-to Memorial Library, Hong Kong).

Hoadley, J. S., *The Government and Politics of Hong Kong: A Descriptive Study with Special Reference to the Analytical Framework of Gabriel Almond* (unpublished University of California, Santa Barbara, Ph.D. thesis, 1968) (copy at the Hung On-to Memorial Library, Hong Kong).

Hong Kong and China: Influence and Interaction Seminar (1981) (a collection of seminar papers) (Hung On-to Memorial Library, Hong Kong).

The Hong Kong Directory 1949 (Hong Kong, The Newspaper Enterprise Ltd., 1949).

Hong Kong General Chamber of Commerce, *Report for the Period 1941–1946* (Hong Kong, Ye Olde Printerie Ltd., 1947).

——, *Report for the Year 1947* (Hong Kong, Ye Olde Printerie Ltd., 1948).

——, *Report for the Year 1948* (Hong Kong, Ye Olde Printerie Ltd., 1949).

——, *Report for the Year 1949* (Hong Kong, Ye Olde Printerie Ltd., 1950).

——, *Report for the Year 1950* (Hong Kong, Ye Olde Printerie Ltd., 1951).

——, *Report for the Year 1951* (Hong Kong, Ye Olde Printerie Ltd., 1952).

——, *Report for the Year 1952* (Hong Kong, Ye Olde Printerie Ltd., 1953).

Ingrams, Harold, *Hong Kong* (London, HMSO, 1952).

Jao, Y. C., Leung, C. K., Wesley-Smith, P., and Wong, S. L. (eds.), *Hong Kong and 1997, Strategies for the Future* (Hong Kong, Centre of Asian Studies, University of Hong Kong, 1985).

Jeffries, Sir Charles, *The Colonial Empire and its Civil Service* (Cambridge, Cambridge University Press, 1938).

——, *The Colonial Office* (London, Allen and Unwin, 1956).

——, *Transfer of Power* (London, Pall Mall Press, 1960).

——, *Whitehall and the Colonial Service: An Administrative Memoir, 1939–1956* (London, The Athlone Press, 1972).

Knaplund, Paul, *British Commonwealth and Empire: 1901–1955* (London, Hamish Hamilton, 1956).

Lau, S. K., *Society and Politics in Hong Kong* (Hong Kong, The Chinese University Press, 1982).

Lee, J. H., *A Half Century of Memories* (Hong Kong, undated, around late 1960s).

Lee, J. M., *Colonial Development and Good Government* (Oxford, Oxford University Press, 1967).

—— and Petter, Martin, *The Colonial Office, War and Development Policy* (London, Maurice Temple Smith for the Institute of Commonwealth Studies, 1982).

Lethbridge, H. J., *Hong Kong: Stability and Change* (Hong Kong, Oxford University Press, 1978).

Lindsay, Oliver, *The Lasting Honour* (London, Hamish Hamilton, 1978).

Lloyd, T. O., *Empire to Welfare State* (Oxford, Oxford University Press, 1979, second edition).

Lo, Hsiang-lin, *Hong Kong and Western Cultures* (Honolulu, East-West Center Press, 1963).

Lorenzo, M. K., *The Attitude of Communist China Towards Hong*

Kong (unpublished University of Chicago A. M. dissertation, 1959) (copy at Hung On-to Memorial Library, Hong Kong).

Louis, W. R., *Imperialism at Bay* (Oxford, Oxford University Press, 1977).

Luard, Evan, *Britain and China* (Baltimore, The John Hopkins Press, 1962).

Luff, John, *The Hidden Years* (Hong Kong, South China Morning Post, 1967).

——, *Hong Kong Cavalcade* (Hong Kong, South China Morning Post, 1978).

Luzzatto, R. and Walker, J. (ed.), *Hong Kong Who's Who* (Hong Kong, Luzzatto and Walker, undated, around 1971).

Madden, A. F. and Fieldhouse, D. K., *Oxford and the Idea of Commonwealth* (London, Croom Helm, 1982).

Mander, John, *Great Britain or Little England* (London, Secker and Warburg, 1963).

Mattocks, K., *This is Hong Kong: The Story of Government House* (Hong Kong, Government Information Services, 1978).

McIntyre, W. D., *The Commonwealth of Nations: Origins and Impact, 1869–1971* (Minneapolis, University of Minnesota Press, 1977).

Miller, J. D. B., *The Commonwealth in the World* (London, Gerald Duckworth & Co. Ltd., 1965).

Mills, L. A., *British Rule in Eastern Asia* (London, Humphrey Milford, 1942).

Miners, N. J., *The Government and Politics of Hong Kong* (Hong Kong, Oxford University Press, 1975).

Morgan, D. J., *The Official History of Colonial Development*, Vol. 5 (London, Macmillan & Co., 1980).

Morgan, W. P., *Triad Societies in Hong Kong* (Hong Kong, Government Printer, 1960).

Morris, James, *Farewell the Trumpets* (Harmondsworth, Penguin Books, 1981).

Morris-Jones, W. H. and Fischer, G. (ed.), *Decolonisation and After* (London, Frank Cass, 1980).

Mushkat, Miron, *The Making of the Hong Kong Administrative Class* (Hong Kong, Centre of Asian Studies, University of Hong Kong, 1982).

Olver, A. S. B., *Outline of British Policy in East and Southeast Asia: 1945–1950* (London, Royal Institute of International Affairs, 1950).

Ommanney, F. D., *Fragrant Harbour* (London, Hutchison, 1962).

Osgood, Cornelius, *The Chinese: A Study of a Hong Kong Community* (Tucson, Arizona, University of Arizona Press, 1975, 3 vols.).

Pearce, R. D., *The Turning Point in Africa: British Colonial Policy, 1938–1948* (London, Frank Cass, 1982).

Pennell, W. V., *History of the Hong Kong General Chamber of Commerce: 1861–1961* (Hong Kong, Cathay Press, 1961).

Pepper, Suzanne, *Civil War in China* (Berkeley, University of California Press, 1978).

Perham, M. F. D., *The Colonial Reckoning* (Connecticut, Greenwood Press, 1976).

Perleberg, Max, *Who's Who in Modern China* (Hong Kong, Ye Olde Printerie Ltd., 1954).

Pope-Hennessy, James, *Half-Crown Colony* (London, Jonathan Cape, 1969).

Potter, B. E., *Britain and the Rise of Communist China* (Oxford, Oxford University Press, 1967).

Rabushka, Alvin, *Hong Kong: A Study in Economic Freedom* (Chicago, University of Chicago Press, 1979).

——, *The Changing Face of Hong Kong* (Stanford, Stanford University Press, 1973).

Rampersad, D. G. M., *Colonial Economic Development and Social Welfare* (unpublished Oxford D.Phil. thesis, 1979).

Rand, Christopher, *Hong Kong: The Island Between* (New York, Alfred A. Knopf, 1952).

Rankin, K. L., *China Assignment* (Seattle, University of Washington Press, 1964).

Reform Club of Hong Kong, *Memorandum and Articles of Association of the Reform Club of Hong Kong* (Hong Kong, Yau Sang Printing Press, 1949).

——, *Tenth Anniversary 1949–1959* (Hong Kong, The Reform Club of Hong Kong, 1960).

Ride, Edwin, *British Army Aid Group* (Hong Kong, Oxford University Press, 1981).

Robinson, K. E., *The Dilemmas of Trusteeship* (London, Oxford University Press, 1965).

—— and Madden, A. F., *Essays in Imperial Government, Presented to Margery Perham* (Oxford, Basil Blackwell, 1963).

Rosecrance, R. N., *Defence of the Realm* (London, Columbia University Press, 1968).

Selwyn-Clarke, Sir Selwyn, *Footprints* (Hong Kong, Sino-American Publishing Co., 1975).

Shively, S., *Political Orientations in Hong Kong* (Hong Kong, Social Research Centre, The Chinese University of Hong Kong, 1972).

Solomon, R. H., *Mao's Revolution and the Chinese Political Culture* (Berkeley, University of California Press, 1971).

Strachey, John, *The End of Empire* (London, Victor Gollancz, 1959).

Strang, Lord, *The Foreign Office* (London, Allen and Unwin, 1955).

——, *The Diplomatic Career* (London, André Deutsch, 1962).

Symonds, Richard, *The British and their Successors* (London, Faber & Faber, 1966).

Tang, Chi-ching, *Social Leaders in Hong Kong and Macau* (in Chinese and English) (Hong Kong, Hong Kong Associated Press, 1958).

Thorne, Christopher, *Allies of a Kind* (London, Hamish Hamilton, 1978).

Tien, Hung-mao, *Government and Politics in Kuomintang China 1927–1937* (Stanford, Stanford University Press, 1972).

Turnbull, C. M., *A History of Singapore, 1819–1975* (Kuala Lumpur, Oxford University Press, 1977).

Union Research Institute, *Who's Who in Communist China* (Hong Kong, the Institute, 1966).

Walden, John, *Excellency, Your Gap is Showing* (Hong Kong, Corporate Communications Ltd., 1983).

Ward, R. S., *Asia for the Asiatics?* (Chicago, University of Chicago Press, 1945).

Webb, D. S., *Hong Kong* (Singapore, Eastern University Press, 1961).

Wesley-Smith, Peter, *Unequal Treaty, 1898–1997* (Hong Kong, Oxford University Press, 1980).

Whiting, A. S., *China Crosses the Yalu* (Stanford, Stanford University Press, 1960).

Wight, Martin, *The Development of the Legislative Council, 1606–1945* (London, Faber & Faber, 1946).

2. Articles in English

Chan, Lau Kit-ching, 'The Hong Kong Question during the Pacific War (1941–1945)', *Journal of Imperial and Commonwealth History*, Vol. 2, 1973–4, pp. 56–78.

——, 'The United States and the Question of Hong Kong, 1941–1945', *Journal of the Hong Kong Branch of the Royal Asiatic Society*, Vol. 19, 1979, pp. 1–20.

Davies, S. N. G., 'One Brand of Politics Rekindled', *Hong Kong Law Journal*, Vol. 7, 1977, pp. 44–80.

De Almada E Castro, Leo, 'Some Notes on the Portuguese in Hong Kong', *Institvio Portvgves De Hong Kong Boletim*, No. 2, September 1949, pp. 215–76.

Department of External Affairs (Australia), 'Hong Kong', *Current Notes on International Affairs*, Vol. 20, 1949, pp. 679–85.

Department of External Affairs (New Zealand), 'Hong Kong', *External Affairs Review*, Vol. 4, 1954, pp. 34–42.

Diamond, A. I., *The Public Records Office of Hong Kong*

(unpublished paper deposited at Hung On-to Memorial Library, Hong Kong).

Endacott, G. B., 'A Hong Kong Symposium', *Journal of Oriental Studies*, Vol. 1, 1954, pp. 324–79.

——, 'Public Administration in Hong Kong: A Review of Sir Charles Collins' Book', *Journal of Oriental Studies*, Vol. 1, 1954, pp. 225–30.

——, 'Proposals for Municipal Government in Early Hong Kong', *Journal of Oriental Studies*, Vol. 3, 1956, pp. 75–82.

Fifield, R. H., 'Hong Kong: Symbol of the West', *International Journal*, Vol. 5, 1950, pp. 230–7.

Freedman, Maurice, 'Shifts of Power in the Hong Kong New Territories', *Journal of Asian and African Studies*, Vol. 1, 1966, pp. 3–12.

Gibson, Tony, 'Hong Kong and the Chinese Opposition', *Eastern World*, Vol. 2, 1948, pp. 10–11.

Grantham, A. W. G. H., 'Hong Kong', *Journal of the Royal Central Asian Society*, Vol. 44, 1959, pp. 119–29.

——, 'Housing Hong Kong's 600,000 Homeless', *Geographical Magazine*, Vol. 31, 1959, pp. 573-86.

Hall, H. D., 'The British Commonwealth as a Great Power', *Foreign Affairs*, Vol. 23, 1945, pp. 594–608.

Hanna, W. A., 'Communist Challenge to British Hong Kong', *East Asia Series*, Vol. 16, 1969, pp. 1–15.

Harcourt, C. H. J., 'The Military Administration of Hong Kong', *Journal of the Royal Central Asian Society*, Vol. 34, 1947, pp. 7–18.

Harris, P. B., 'Representative Politics in a British Dependency: Some Reflections on Problems of Representation in Hong Kong', *Parliamentary Affairs*, Vol. 28, 1975, pp. 180–98.

Hinton, W. J., 'Hong Kong's Place in the British Empire', *Journal of the Royal Central Asian Society*, Vol. 28, 1941, pp. 256–69.

Hoadley, J. S., 'Political Participation of Hong Kong Chinese: Patterns and Trends', *Asian Survey*, Vol. 13, 1973, pp. 604–16.

'Hong Kong', *Current Affairs Bulletin*, Vol. 9, 1951, pp. 51–63.

Ingrams, Harold, 'Hong Kong and its Place in the Far East', *Journal of the Royal Central Asian Society*, Vol. 40, 1953, pp. 255–65.

Kirby, E. S., 'Hong Kong Looks Ahead', *Pacific Affairs*, Vol. 22, 1949, pp. 173–8.

Larsen, H. W., 'Impressions of Postwar Currency Problems in Hong Kong', *Pacific Affairs*, Vol. 19, 1946, pp. 285–90.

Lau S. K., 'Social Change, Bureaucratic Rule and Emergent Political Issues in Hong Kong', *World Politics*, Vol. 35, No. 4 (1983), pp. 544–62.

————, 'The Government, Intermediate Organizations, and Grass-roots Politics in Hong Kong', *Asian Survey*, Vol. 21, No. 8 (1981), pp. 865–84.

Lowe, C. J. G., 'How the Government in Hong Kong Makes Policy', *Hong Kong Journal of Public Administration*, Vol. 2, 1980, pp. 63–70.

Madden, A. F., 'How Colonies Grow into Dominions', *United Empire*, Vol. 41, No. 3, 1950, pp. 159–64.

Pennell, W. V., 'Hong Kong Looks to a Greater Future', *Canadian Geographical Journal*, Vol. 47, 1953, pp. 2–13.

Podmore, David, 'Localization in the Hong Kong Government Service, 1948–1968', *Journal of Commonwealth Political Studies*, Vol. 9, 1971, pp. 36–51.

Sung, Kayser, 'China's Shadow over Hong Kong', *Contemporary China*, 1968, pp. 98–112.

Stewart, G. O. W., 'Post–war Development in Hong Kong', *Asian Review*, Vol. 58, 1962, pp. 128–31.

Wolf, D.C., '"To Secure a Convenience": Britain Recognizes China — 1950', *Journal of Contemporary History*, Vol. 18, 1983, pp. 299–326.

Woodhead, H. G. W., 'Shanghai and Hong Kong', *Foreign Affairs*, Vol. 23, 1945, pp. 295–307.

3. Books in Chinese

Chan, Wai-wang (陳偉宏), *Hong Kong, Taipei, Peking: A Trilateral Relationship* (香港，台北，北京：三國之間的微妙關係) (original English subtitle) (Hong Kong, International Affairs College Press, 1980).

Chang, Sheng (章盛), *Hsiang-chiang heh-shê-hui huo-tung cheñ-hsiang (The Real Story of the Activities of the Triad Societies in Hong Kong* (香港黑社會活動真相) (Hong Kong, Cosmos Books Limited, 1983).

Cheung, H. T. and Lo, T. K. (張漢德與盧子健), *Political Reform in Hong Kong – The Past and the Future* (政制改革：何去何從) (original English subtitle) (Hong Kong, Genius Publishing Company, 1984).

Chin Jun (錦潤), *Chiang-tu lien-ch'uan (Biographical Sketches of Hong Kong Governors)* (港督列傳) (Hong Kong, 1983).

Ch'un-chung ch'u-pan-she (群衆出版社), *Mei chiang t'e wu tsai chiang chiu ti tsui hsing (The Atrocious Activities of the American and Kuomintang Spies in Hong Kong)* (美蔣特務在港九的罪行) (Peking, 1957).

Hsiang-chiang yu chung-kuo li shih wen hsin tsu liao hui pien (香港與中國歷史文獻資料彙編) (Hong Kong, Wide Angle Press Ltd., 1981).

Hsin-wen-yen-chiu tsu-liao tsung-ti-shih-erh-chi (新聞研究資料總
第十二輯) (Peking, Chan-wang chu-pan-she, 1982).

Li, Chin-wei (黎晉偉), *Centenary History of Hong Kong* (香港百年史)
(original English subtitle) (Hong Kong, 1948).

Lin, Yu-lan (林友蘭), *Hsiang-chiang shih-hua* (香港史話) (Hong
Kong, Shanghai Book Store, 1980).

Lu, Yen (魯言), *Hsiang-chiang chiang-ku (Hong Kong Anecdotes)*
(香港掌故) (Hong Kong, Wide Angle Press Ltd., 1977–83, 6
vols.).

Wah kiu yat pao (華僑日報), *Hsiang-chiang nien-chien (1948–1953)*
(香港年鑑) (Hong Kong, Wah kiu yat pao, 1948–53).

Wang, Ching (王敬), *Hsiang-chiang i wan fu hao lieh ch'uan
(Biographical Sketches of the Billionaires in Hong Kong)*
(香港億萬富豪列傳) (Taipei, 1980).

Wu, K. P. (吳健平), *The Story of the Anti-Communistic Activities of
the Patriotic Refugees now at Rennie's Mill Camp, Hong Kong*
(調景嶺義民反共奮鬥史實) (original English subtitle) (Taipei,
Universal Voice Press, 1958).

Yong, T. F. (楊德和) (ed.), *A Chronology of Hong Kong in the Last
Decade: 1944–1954* (香港十年大事記 1944–1954) (original
English subtitle) (Hong Kong, Johnson Brothers, 1954).

4. Articles in Chinese

Chi (吉), 'Te wu tao liao hsiang-chiang (特務到了香港) (The Spies
Have Arrived in Hong Kong)', *Kwang ming pao (hsun kan)*
(光明報 (旬刊)) No. 1 (Hong Kong, 1946).

Sun, Po I (孫寶毅), 'Hsiang-chiang wen hua ti fen hsi
(香港文化的分析) (An Analysis of the Hong Kong Culture)', *The
Author's News* (作者通訊) (original English subtitle), Vol. 1
(Hong Kong, 20 April 1951).

Index

on reform, 154, 166
Stanley, Lord, 2
Stanley, Oliver, 12, 21
Stephen, Sir James, 2
Stuart, J. L., 46
Sum Chit-son, 29
Sun Fo, 54–5
Swire, G. W., 18

TA KUNG PAO, 134, 140, 178, 206; circulation in 1950, 132
Taiwan, 159, 181
Tanganyika, 188
Tat Tak Institute, 85–6, 125, 135
Tobago, 188
Todd, (Revd) Alastair, 188
Tramway strike (1949–50), 126–30, 206
Treasury, 46, 48, 72–3, 157, 195; attitude towards reform, 43, 65–6; ending of control over Hong Kong, 201
Treaties: First Convention of Peking (1860), 1; Second Convention of Peking (1898), 1, 81; Treaty of Nanking (1843), 1
Trench, Sir David C. C., 213
Triad, 125, 137, 182
Trinidad, 188
Tsui Tung-loi (General), guerrilla activities of, 137

UNIONS: DANGER OF POLITICAL INFILTRATION OF, 206–7; Hong Kong Federation of Trade Unions, 88, 127–30, 139, 178, 181, 206–7; Hong Kong Government's policy towards, 87–9; Hong Kong and Kowloon Trade Union Council, 88–9, 139, 206–7; Kuomintang control of, 43, 51–2, 85; Tramways Union, 127–9, 206
United States: threat to cut off Marshall Aid to Britain, 131; trade restrictions on Hong Kong, 169–70; vote of no-confidence in Hong Kong's future, 172

University of Hong Kong, 35
Unofficial Justices of the Peace, 3–5, 35, 111, 115, 208, 214
Urban Council, 17–18, 34, 39, 75, 93–4, 97–8, 100, 108, 115, 122, 164–8, 208; composition before the Second World War, 5; first elections after the Second World War, 67–8, 158–61, 204–5; Grantham's proposed changes of, 110–16, 161, 166–8, 186; own proposals for reform, 164–5; re-establishment of (1946), 47

WAH KIU YAT PAO, 172; circulation in 1950, 132; Kuomintang attempt to capture, 50–1
Wallace, W. I. J., 65, 193
Wallinger, G. A., 29
Wang Shih-chieh (Dr), 83
Watson, M. M., 196; attitude towards reform, 96–7
Wen hui pao, 140; circulation in 1950, 132
Whitehead, T. H., 3
Wong Chung-man (General), guerrilla activities of, 137
Wong San-yin (Dr), 145
Woo, P. C., 159–60
Wu Te-chen (General), 29, 84

YOUNG, SIR MARK A., 67, 110–11, 196–7; comparing his views with Grantham's on reform, 186–90; early views on reform, 23; handling of 1946 Kowloon Walled City incident, 82; reform proposals (the Young Plan), 32–47, 63–5, 70, 72–4, 77, 80, 89, 91, 93, 95–9, 105, 108–9, 121, 138, 140, 143–4, 153, 183–5, 189, 193, 198–200, 202, 207–10; restoration of civil government (1946), 47–8; views on Hong Kong's future, 186–7, 189–90
Yung Hao, SS, Hong Kong acquisition of, 175

Note: In the Index, personal titles are given in parentheses if the person concerned was not titled during the period covered by this book.